Magical Haskell

A Friendly Approach to Modern Functional Programming, Type Theory, and Artificial Intelligence

Anton Antich

Apress®

Magical Haskell: A Friendly Approach to Modern Functional Programming, Type Theory, and Artificial Intelligence

Anton Antich
Schaffhausen, Schaffhausen, Switzerland

ISBN-13 (pbk): 979-8-8688-1281-1   ISBN-13 (electronic): 979-8-8688-1282-8
https://doi.org/10.1007/979-8-8688-1282-8

Copyright © 2025 by Anton Antich

This work is subject to copyright. All rights are reserved by the Publisher, whether the whole or part of the material is concerned, specifically the rights of translation, reprinting, reuse of illustrations, recitation, broadcasting, reproduction on microfilms or in any other physical way, and transmission or information storage and retrieval, electronic adaptation, computer software, or by similar or dissimilar methodology now known or hereafter developed.

Trademarked names, logos, and images may appear in this book. Rather than use a trademark symbol with every occurrence of a trademarked name, logo, or image we use the names, logos, and images only in an editorial fashion and to the benefit of the trademark owner, with no intention of infringement of the trademark.

The use in this publication of trade names, trademarks, service marks, and similar terms, even if they are not identified as such, is not to be taken as an expression of opinion as to whether or not they are subject to proprietary rights.

While the advice and information in this book are believed to be true and accurate at the date of publication, neither the authors nor the editors nor the publisher can accept any legal responsibility for any errors or omissions that may be made. The publisher makes no warranty, express or implied, with respect to the material contained herein.

Managing Director, Apress Media LLC: Welmoed Spahr
Acquisitions Editor: Melissa Duffy
Development Editor: James Markham
Coordinating Editor: Gryffin Winkler

Cover designed by eStudioCalamar

Distributed to the book trade worldwide by Springer Science+Business Media New York, 1 New York Plaza, New York, NY 10004. Phone 1-800-SPRINGER, fax (201) 348-4505, e-mail orders-ny@springer-sbm.com, or visit www.springeronline.com. Apress Media, LLC is a Delaware LLC and the sole member (owner) is Springer Science + Business Media Finance Inc (SSBM Finance Inc). SSBM Finance Inc is a **Delaware** corporation.

For information on translations, please e-mail booktranslations@springernature.com; for reprint, paperback, or audio rights, please e-mail bookpermissions@springernature.com.

Apress titles may be purchased in bulk for academic, corporate, or promotional use. eBook versions and licenses are also available for most titles. For more information, reference our Print and eBook Bulk Sales web page at http://www.apress.com/bulk-sales.

Any source code or other supplementary material referenced by the author in this book is available to readers on GitHub (https://github.com/Apress). For more detailed information, please visit https://www.apress.com/gp/services/source-code.

If disposing of this product, please recycle the paper

To my parents for making me curious and my wife for her patience, support, and inspiration while working on this book.

Table of Contents

About the Author xi

About the Technical Reviewer xiii

Acknowledgments xv

Introduction xvii

How This Book Came to Be: Instead of an Introduction xix

Chapter 1: Wizards, Types, and Functions 1
    Solving Problems As Wizards Do... 3
    Let's Fall in Love with Types! 7
    A Char and an Int Walk into a Bar and Make a Function 8
    How FunTypical!.. 12
    Curry and Recurse!... 15
    Conclusion 19

Chapter 2: Type Construction 21
    House of Cards with a Little Help from Algebra 22
    Algebraic Data Types.. 25
    Records 27
    Type Functions Are Functions, Maybe?... 30
    Advanced: On Data Constructors and Types... 37

TABLE OF CONTENTS

    List, Recursion on Types and Patterns 39
        Length of a List 42
        Map a Function Over a List of Values 44
    Conclusion 50

Chapter 3: Very Gentle Type Theory and Category Theory Intro 51
    Types and Functions 52
    Maybe and Advanced Generalized Functions 54
    Dependent Function Types (Pi-Types) 59
    Sum, Product, and Dependent Pair (Sigma) Types 61
    Very Gentle Category Theory Introduction 63
    Typeclasses 66
    Conclusion 70

Chapter 4: Basic Typeclasses or "Show Me a Monoid" 71
    Show Typeclass 72
    Algebra Is Cool 73
        Typeclass Hierarchy in Haskell 76
    Lift Me Up! 79
    Conclusion 83

Chapter 5: Functor, Bifunctor, and Applicative Functor Enter an Elevator 85
    Functor Typeclass Definition 86
    Three-Dimensional Vector Example 89
    Tracking Players in Our Game 91
    You Are Either Functor or a Bifunctor 92
        We Need a Bigger Lift 98

Action! Apply! Cut! 104
Applicative Typeclass 107
Conclusion 111

Chapter 6: O, Monad, Help Me Compose! 113
Can We Play Cards in This State? 117
Monad Typeclass and Basic Monads 128
Reader–Writer–State Triple 132
The State Monad 132
The Reader Monad 135
The Writer Monad 140
Conclusion 145

Chapter 7: Input, Transformer Stack, Output 147
Do Notation 153
Monad Transformer Stacks 158
Conclusion 169

Chapter 8: Blackjack: Full Haskell Program 171
Preparation 174
Initial "Pure" Design 177
Building the First Floor: IO 183
Building the Top Floor: StateT IO 188
Final Game and Recap 197
Conclusion 201

Chapter 9: Let's AI 203
Agentic AI and Large Language Models 203
LLMs in a Nutshell 205

TABLE OF CONTENTS

Haskell and LLMs 210
    Big Picture of the Haskell AI Framework 211
    Terminal Chatbot 212
Conclusion 231
Conclusion 234

Chapter 10: Terminal AI Chat Agent 237

Terminal UI Skeleton 238
    Adding a Nicer Response Function 239
    Initialization 241
    Building the Monad Transformer Stack 242
    Main Program 245
Building Muscles 248
    Logging 248
    Usage Stats in the Writer Monad 253
    Chat History 255
Conclusion 263

Chapter 11: Web-Enabled AI Framework and GHC "Guts" 265

Terminal UI Improvements 265
    Example Jarvis Scenarios 270
"M" Subchapter: Under the Hood of GHC 278
Adding a Web API 283
Mutable Variables 294
    IORef: Simple Mutable Variables in the IO Monad 295
    MVar: Mutable Variables with Concurrency Control 296
    TVar: The Power of Software Transactional Memory (STM) 298
Conclusion 300

Chapter 12: Down the Rabbit Hole 303

MongoDB as a Persistence Layer 304

MongoDB Initialization 308

Adding a System Prompts Collection 310

Redesigning Architecture: MRWST Monad Transformer 313

Vectors and Type Families 321

Type Synonym Families 322

Data Families 323

Associated Type and Data Families 325

Vectors As Arrays in Haskell 330

More AI: Vector RAG 340

In-Memory Vector Engine 344

Conclusion 358

Chapter 13: AI Multi-agents, Arrows, and the Future 359

Arrows: The Last Abstraction 361

Arrows Definition and Basic Examples 364

Arrows Interface 367

AI Multi-agents and Arrows 373

AI Multi-agents Examples 377

Arrows to Support Quick AI Multi-agent Creation 379

Final Improvements for the AI Framework 390

Web Authentication with JWTs 390

Error Handling: Embracing the Try–Catch Approach 396

Multiple LLMs and Agents Support 399

Conclusion 404

Index **407**

About the Author

Anton Antich is a serial entrepreneur and AI/Type Theory researcher, who majored initially in physics. In various executive roles, he was instrumental in building Veeam Software from 0 to over 1B USD in annual sales in under ten years – making Veeam the fastest-growing European company ever. Upon exiting Veeam, he invested in and helped scale over 20 startups and in the recent years got back to the roots and is currently building Integrail (`https://integrail.ai`) – a no-code Agentic AI platform.

Haskell has been an integral part of many of the projects undertaken during Anton's tenure, and his current startup will use it as an optional backend for AI multi-agent execution.

About the Technical Reviewer

German Gonzalez-Morris is a polyglot software architect/engineer with 20+ years in the field, with knowledge in Java, Spring Boot, C/C++, Julia, Python, Haskell, and JavaScript, among others. He works with cloud (architecture), web-distributed applications, and micro-services. German loves math puzzles (including reading Knuth, proud by solving some of Don's puzzles), swimming, and table tennis. Also, he has reviewed several books, including an application container book (Weblogic) and books on languages (C, Java, Spring, Python, Haskell, TypeScript, WebAssembly, Math for coders, regexp, Julia, Data Structures and Algorithms, Kafka).

Acknowledgments

There are too many people I would like to acknowledge for making this book a reality, and I would like to list at least a few.

My aunt Yulia for making me fall in love with physics at the age of six – and without physics there would be no journey to Haskell for me – I do hope you are in the better world now.

Richard Feynman for inspiring me to be a better scientist even without formally being one.

All the amazing people working on making Haskell, and GHC specifically, the cutting edge of computer science.

Ratmir Timashev for everything I've learned doing business and for making Agentic AI at Integrail a reality.

My grandfather for giving me an amazing C book when all I knew was Pascal.

My parents for making me curious and giving me the "programmable calculator MK-61" 40 years ago, the programming platform before all others.

My lovely wife, Anastasia, for supporting me in everything I do and for the patience with me while working on this book.

The Creator of the Universe for making all of it possible.

Introduction

Since I was six years old, I wanted to be a nuclear physicist. The idea of being captivated by the *beauty* of how the world works, expressed in the language of mathematics, fascinated me. Later in life it transformed to being captivated by the beauty of mathematics itself regardless of applications – physics, AI, modeling of advanced business processes in sales, or election results. I studied physics as planned; did research for some time; and learned a bunch of programming languages starting with Pascal and Fortran via old books in my father's library and moving to C, then to C++, and finally to Java, as it became arguably the most fashionable language in the 1990s, and used all of them professionally in various capacities. Then, I switched to the business side of things and stopped programming, but always kept math and research as a "hobby" of sorts.

Then, I discovered Haskell.

Oh, how I wish that I'd discovered it sooner. No other *practical* language suitable for the creation of high-performance production apps comes closer to pure math than Haskell. No other language gives you goosebumps from the code you write. Discovering it led to a bumpy journey toward Type Theory and modest Category Theory beginnings, where the beauty of math combined with the ability to write code resulted in something beautiful *and* practical. This journey took about ten years, with lots of frustrations along the way. The experience of this journey allowed me to formulate hopefully a somewhat different approach to teaching Haskell than what has been available so far – and delivering it is the purpose of this book.

INTRODUCTION

As I was working on it, an AI revolution happened and I started an Agentic AI company along the way – thus, the second half of this book is fully dedicated to Agentic AI and approaches to implementing it in Haskell.

This book is for anyone who is interested in learning Haskell, maybe made a couple of attempts and got frustrated by not understanding how to combine all the types into the working program or what monads are. I hope some of the pitfalls I fell into while learning you will be able to avoid using the carefully-built-from-the-ground-up knowledge tree of types, functions, typeclasses, and other abstractions, introduced gently and with practical examples.

And even if you do not use Haskell in production anytime soon, learning these principles will make you a better programmer in any other language.

I hope you will enjoy it as much as I loved working on it.

How This Book Came to Be: Instead of an Introduction

Feel free to skip this somewhat philosophical intro right to Chapter 1.

Lately, I've been spending quite a bit of time with Type Theory, lambda cube (e.g., implementing System F-omega-ish type library in CoffeeScript for the fun of finding out how things work as Dr. Feynman used to say), toy functional languages (Haskell is great, but what would a great functional language look like if it were designed today – with all that we've learned from state-of-the-art Haskell development in the last 20 years?), parsers, GUI libraries design, and functional reactive programming.

Three thoughts persisted:

"I need to write another monad tutorial" – haha, got ya. "How do I teach functional programming to my kids?" – so that it's light, comprehensible, fun and conveys the beauty of the concepts involved. And "we've been doing it all wrong."

Okay, the latter may be an exaggeration, but I need to drive the point across. Like many unfortunate souls before me, I have gone down the painful but oh-so-typical road of coming from an imperative object-oriented background and trying to build bridges from the patterns learned there to a completely, absolutely different functional world.

The problem with this is, your brain works against you (*do read the classic Thinking, Fast and Slow (*https://www.amazon.com/dp/B00555X80A/*) by Dr. Kahneman if you haven't yet – at the very least, it will help you detect and fight your own biases*).

HOW THIS BOOK CAME TO BE: INSTEAD OF AN INTRODUCTION

Our brain's "System 1" works in patterns. It is efficient, extremely fast – much faster than the conscious "System 2," which we are using when studying Category Theory – it is in contrast subconscious, and it deconstructs and shreds all the elephants you see in your life into basic shapes and curves and colors, which makes recognizing elephants a very easy task, until one day you glimpse upon a fluffy cloud and your brain goes: "Oh, look, it's an elephant!"

Well, no, it's a cloud.

And this is exactly the trap laid out for programmers trying to escape to the wonderful world of Haskell from the less elegant Java, C++, or those *terrible-which-shall-not-be-named-but-end-in-...avaScript* backgrounds. The thought goes: C++ is a programming language, as is Haskell, so the concepts should be pretty similar, right? Isn't it like Indian and African elephants? Sure, one has bigger ears, but aren't they essentially the same?

No, they are not: one is an elephant, and the other one is a cloud. You cannot reason about clouds using your elephant knowledge. You can ride an elephant, but attempting to ride a cloud would end really badly. Your brain constantly tries to find analogies between your previous experience and new knowledge – that's how we learn, it is a good thing – but in this case the unfortunate fact that both Haskell and JavaScript are called "programming languages" is a disservice, as trying to find similar patterns in both leads to consequences only slightly less disastrous than trying to ride a cloud thinking it's an elephant.

To *master* the power of Haskell, you need to set aside everything you know about imperative programming and build upon a totally different foundation. One of pure abstract math, starting from Category Theory, from set theory, from Type Theory, from Henri Poincare. You have to build natural numbers with pen and paper starting with an empty set, construct a Ring on them, abstract away and discover Monoids, start appreciating the beauty of the internal structure of types and categories manifested in polymorphic functions and laws on them, be amazed at how one abstraction can elegantly describe seemingly unrelated entities ...

HOW THIS BOOK CAME TO BE: INSTEAD OF AN INTRODUCTION

But who in the world has time for this?!

Well, exactly. This is one of the primary reasons there's still a certain bias against Haskell among practitioners – a big part of the Haskell community give the impression of very smart monad-stacking-arrow-composing-scientist-magicians sitting in their ivory towers with too much time on their hands working out some abstract problems, while *we need to code stuff that works even if it's ugly.*

I have been using Haskell both in production and hobbyist projects for over 15 years now. Currently we are building an AI agents platform using Haskell as an integral part, so what I would like to say is: take heart, there's hope!

But before I outline a proposed (yet another) way out, let me illustrate the elephant vs. cloud conundrum with a pretty typical "Where is my `for` loop?!" situation.

Let's say you are building a GUI library based on some low-level API and need to render lines of text on the screen. You construct an array of textures where each corresponds to a new line of text and need to render each one under the other, incrementing a y coordinate, like so:

```
int y = 100;
int step = 40;
for (int i = 0; i < numOfTextures; i++) {
  renderTexture (texture[i], y);
  y += step;
}
```

Something you've done a thousand times in different situations. You switch to Haskell where you have a list of textures `textures :: [Texture]`, and your System 1 starts screaming *"Where is my `for` loop?!"* and then *"How do I pass the state when iterating?"* Should I use mutable variables, `IORef`, `MVar`, `TVar`? I know I am supposed to map over lists, so should I use a `State` monad to increment y while mapping?

HOW THIS BOOK CAME TO BE: INSTEAD OF AN INTRODUCTION

And only when you tell your System 1 to stop getting ideas from your imperative experience, consider abstractions, realize it is best to *fold* and *traverse* structures in the functional world, you come to something like

```
foldM (iterate step) startingY textures
  where iterate s y tex = renderTexture tex y >> (y + s)
```

...which gives you goosebumps after all the braces and semicolons in the C-land, but patterns have nothing in common: *you iterate over arrays with mutable variables* in one, and you *fold structures with computations* in the other. Elephants and clouds.

So how do we teach Haskell to kids or help adults master its power faster and more efficiently?

First, stop trying to build bridges from the imperative world – it messes with people's brains. In fact, explaining design patterns for program architecture on the real-world problems and *contrasting* approaches with the imperative world is very much needed and largely missing now. However, this needs to be done only after a student has learned the functional approach and small patterns from scratch. Teaching Haskell in the bottom-up imperative way, with "Hello World" and calculator in ghci, is in my strong opinion absolutely counterproductive – it leads right into the elephants vs. clouds trap and takes much longer for the student to start appreciating the power and beauty of Haskell.

We need to start from Type Theory, make people understand and fall in love with types, their structure, their daily lives, their relations with each other; a student has to become fluent in writing and understanding type-level functions before they even think about writing "real" functions or "Hello World" for that matter. It is actually a pretty traveled road as well – but onlyif you have a reasonably strong math background.

Catch up on your Category Theory reading (Bartosz Milewski's *Category Theory for Programmers* is out and it's excellent), study Typeclassopedia (`https://wiki.haskell.org/Typeclassopedia`), use the amazing Stephen Diehl's *What I Wish I Knew When Learning Haskell* (http://dev.stephendiehl.com/hask/) as a reference, and get a feeling on how to structure real programs using monad transformer stacks, but whatever you do, under any circumstances, **DO NOT read monad tutorials!**

Monads are very simple, it's just another beautiful and powerful math abstraction, one among many, and you will confuse and break your brain with false analogies if you read the tutorials – study it in the context of Functor–Applicative–Monad hierarchy, so, again, Typeclassopedia, or, of course and hopefully, this book.

Then, what was the point of all this? I am a very visual person when studying, and I've been thinking for some time now that math concepts underlying modern functional programming and Type Theory can be explained in a more accessible way using visuals, magic, and various cute animal-related analogies. Existing literature on the concepts mentioned above is often too technical, even for practicing programmers, let alone kids, even in their teens, and focuses on formal proofs a lot.

However, if we ease on trying to rigorously prove stuff and just illustrate and explain those same concepts, from lambda calculus in SystemFw (*even the name is a mouthful, but the machinery is again extremely simple!*) to types and their relations to Category Theory and arrows, in a top-down visual approach, this may help people build new patterns, which are completely unrelated to imperative programming. Then gradually move to a more technical, more in-depth study while already becoming productive in writing elegant Haskell code for practical tasks. Something in the spirit of excellent "Learn you a Haskell" (learnyouahaskell.com) but building up upon the math foundation of types and typeclass hierarchy vs. "Hello World" and ghci REPL.

Hence, the idea of *Magical Haskell* was born. In fact, it should probably go well hand in hand with *Learn You a Haskell for Great Good!* – with the latter providing for a lot of basic language foundation in much more detail and *Magical Haskell* focusing on trying to create a system of typed functional patterns, typeclass hierarchy, and explaining abstract technical concepts in an accessible comprehensible way.

HOW THIS BOOK CAME TO BE: INSTEAD OF AN INTRODUCTION

Short Introduction: How to Read This Book

In this book, we make an attempt to introduce advanced and at times *very* technical concepts that underlie Haskell language in an accessible way, using everyday analogies, pictures, and real-world examples where possible. We have a strong conviction that having a strong grasp on types and the bigger picture, in general, makes any practical Haskell programmer much more efficient and productive while also being able to fully appreciate the beauty of the mathematical concepts involved. We start with these foundations and then move to the advanced concepts of how to structure real-world big applications (a topic a lot of Haskell students struggle with) using a variety of abstractions, libraries, and design patterns.

We start **Chapter** 1 with basic types and functions, introducing the functional approach to solving problems along the way. **Chapter** 2 continues with an introduction to algebraic data types, parametric types ("type functions"), recursion, and pattern match. We also look at a concise program that deals with a deck of cards and uses a lot of these concepts. At the end of this chapter, we "invent" a **Functor**. **Chapter** 3 focuses on a semiformal introduction to the (Martin-Loef) Type Theory and very lightly touches on Category Theory. Haskell uses a weaker type system called "System FC," but again, understanding the bigger picture helps to both grasp Haskell abstractions and transition to languages such as Idris or Coq with less difficulty.

Chapter 4 focuses on basic **typeclasses** using ubiquitous algebraic examples (such as **Monoid**) to explain the concept. **Chapter** 5 follows up with and focuses on how to introduce **Functor** and **Bifunctor** naturally and finishes with **Applicative** discussion, a further specialization of the Functor. In **Chapter** 6, we gently introduce **Monads**, showing again how they naturally extend the Functor–Applicative–Monad triad, and discuss three most popular monad types: **Reader, Writer, and State**. **Chapter** 7 discusses the **IO Monad**, introduces "do notation," and explains what

Monad transformer stacks are and why they serve as a very good way to structure real-world programs. **Chapter** 8 is the quintessence of all the previous discussion: here we build the simplified "Blackjack" program from scratch step by step, applying all the typed functional design principles that we have learned previously.

Chapter 9 starts the second part of the book, focused on learning increasingly more advanced Haskell concepts along the lines of building an AI framework that will allow us to build AI agents of any complexity. It focuses on an introduction to large language model (LLM)–based AI and building the foundation of the terminal UI–based AI agent – "Jarvis." We design a more sophisticated, production-quality monad transformer stack in this chapter.

Chapter 10 builds out a final working terminal-based Jarvis with OpenAI backend, building "muscles" on the skeleton of the InputT on RWS on the IO monad. We discuss and implement logging and add chat history support for Jarvis.

Chapter 11 adds web API access to our framework using Scotty and discusses the issues we face due to multi-threading of the Web. We peek under the hood of Glasgow Haskell Compiler (GHC) and discuss the reasons behind its efficiency and speed. We also demonstrate some specific use cases where AI Jarvis is useful. We discuss mutability in Haskell.

Chapter 12 is full of advanced practical work. We discover and understand type families and advanced mutability. We learn to use vectors. We build a simple in-memory vector storage that allows us to add RAG (retrieval augmented generation) to Jarvis. We add mongodb as a persistence layer and discover the ST monad and Rank-N polymorphism. Finally, we build a mutable RWS monad from scratch – a very handy trick that allows us to reuse all the previous code in both the web and terminal environments, while MRWS can serve as a standalone library for any web-based applications.

Chapter 13 finishes on the high note, gently introducing the beautiful and powerful "last abstraction" – Arrows. We gradually discover what Arrows are and how they abstract functions, monadic actions, and are able to help us capture much more useful context than monads allow and then build a simplified arrow-based library for AI multi-agent creation. We further extend our AI framework with web security, error handling, and integrating with integrail.ai – a platform that allows us to create various AI multi-agents easy and fast.

Simple Types, Simple Functions, Recursion, If-Else

Download and install Haskell from http://haskell.org or via a package system on your OS. Run ghci in your terminal to work with REPL, or use online REPL at http://repl.it/languages/haskell.

Type names and type constructor names always start with a capital letter.

Functions are defined as

```
-- type signature first: a function converting Int to Int
fact :: Int -> Int
-- recursive definition via if-then-else statement:
fact n = if n == 0 then 1 else n * fact (n-1)
-- no need for parenthesis or commas between function arguments
```

Definition via pattern matching:

```
fact :: Int -> Int
fact 0 = 1
fact n = n * fact (n - 1)
-- patterns are translated into nested case statements under
the hood, in order
```

```
-- fibonacci numbers:
fib :: Int -> Int
fib 0 = 1
fib 1 = 1
fib n = fib (n - 1) + fib (n - 2)
```

Functions of many arguments are defined via *currying* – as a function of one argument that returns a function of one argument that returns a function ...:

```
-- function that takes a Char and an Int and returns an int,
-- is really a function that takes a Char and returns a function
-- that takes an Int and returns an Int
anotherWeirdFunction :: Char -> Int -> Int
anotherWeirdFunction c i = (whichNumberIsChar c) + i
```

```
-- useless 3 argument multiplication function:
mult3 :: Double -> Double -> Double -> Double
mult3 x y z = x * y * z
```

CHAPTER 1

Wizards, Types, and Functions

You wake up in a cheap motel room in Brooklyn. You hear police sirens outside, the ceiling is gray and dirty, and a big cockroach is slowly crawling across. You don't remember much, except that you need to be in a fancy restaurant in Manhattan at 7 PM – oh my, it's in 15 minutes! – and that you are a wizard. The problem is, you only remember three spells (Figure 1-1).

Figure 1-1. *The spells you remember*

CHAPTER 1   WIZARDS, TYPES, AND FUNCTIONS

You are obviously in a hurry, so, looking attentively at a cockroach, you wave your hands and go: "Moz'bah'da!" And puff! There is a beautiful white horse right beside the bed! In a short moment, you realize that riding a horse through New York might not get you there on time, plus you are not sure you even *can* ride a horse, so you wave your hands again and shout: "Gimmeraree!"

The horse disappears in a puff of smoke, and all you get is a pile of stinking goo on the floor. Oh-oh. You forgot that "Gimmeraree" works only on pigs! Poor horse. You catch another cockroach, go outside, and this time say your spells in the correct order – "K'hee-bah!", "Gimmeraree!" – and there you go, a red Ferrari Spider, in mint condition standing right there beside the garbage bins. Hastily, you jump into the car and drive away to a meeting that will hopefully clear up your past, present, and future destiny (see Figure 1-2).

In this book, we will study modern typed functional programming concepts and learn to apply them using Haskell, the most powerful and popular modern functional language. If you have previous imperative programming language exposure (C, C++, C#, Java, JavaScript, etc.), you might be anxious to jump right into writing your first "Hello World" program – we will get there soon enough. For now, I strongly encourage you to put away your imperative baggage, take a step back, and meditate a bit on how we approach solving problems the functional way. Trust me, it will be worth it in the long run: if you learn the concepts, which are much more powerful and elegant, but very different from the imperative world, you will become efficient much quicker – and then you can gradually add your imperative experience into the mix. For now, it is best to pretend it simply does not exist.

Typed functional programming is just like magic. Your spells are functions. They turn something into something else (like pigs to Ferraris or strings to numbers) or produce some effect (like putting ogres to sleep or drawing a circle on the screen), and *you need to combine different spells in the right order to get to the desired result.*

What happened in the story above was *dynamically typed* magic: you used a spell that works only on pigs on a horse instead, and your "program" crashed, killing the horse in the process (that's what often happens with JavaScript programs). With the *statically typed* magic that we will use from now on, a powerful being called **Typechecker** would see your mistake, stop your hand, and tell you that "Gimmeraree" expects a pig as its input – which would indeed make you a much more powerful and successful mage than if you go around killing other people's horses.

Solving Problems As Wizards Do

So how do we approach solving problems the functional way? The mage above needed to get a fast mode of transportation, and all he had were cockroaches. So he had to think about how he could get from a cockroach to a Ferrari using the spells he knows by combining them in the correct order. He started with an end result – Ferrari – and worked backward from there, getting to the correct sequence by *composing* spells that work on the *right* subjects.

Figure 1-2. *Composing spells in the right order, minding the types*

Now, let's get real and say you need to calculate an average of a list of numbers and show it on the screen. Please resist the imperative urge to start *telling the computer what to do and how to do it,* and let's focus on the problem as a wizard would instead.

You have a human, and you need to turn her into a line on the screen, hopefully without killing anyone. Let's work backward from our desired result – "a line on the screen with the average printed out" (Figure 1-3).

Figure 1-3. *Our first problem*

To get to this result, we need something that takes a number and shows it on the screen. Let's call this something "print" (Figure 1-4).

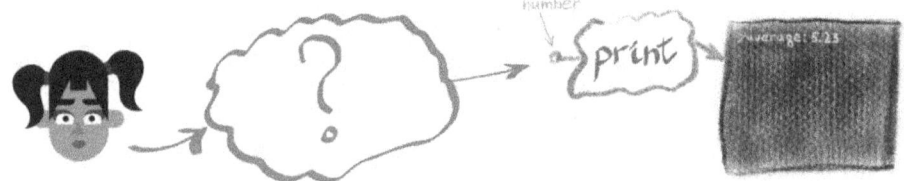

Figure 1-4. *Step 1 in resolving the problem*

Now our problem is a bit easier – we need to get from a human to a number. "Print," whatever it is, needs a number to operate. Since our goal is to calculate and show an average value, the previous step needs to take a list of numbers and, uhm, calculate their average (Figure 1-5).

CHAPTER 1 WIZARDS, TYPES, AND FUNCTIONS

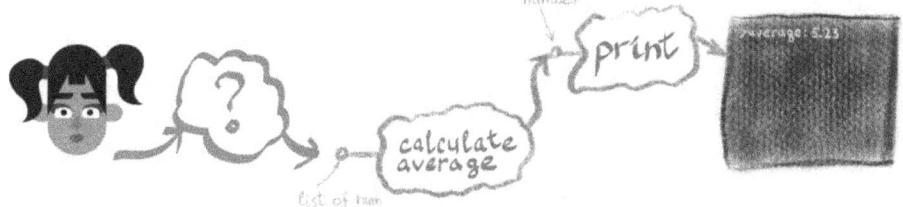

Figure 1-5. *Step 2 in resolving the problem*

Our problem became easier still – now we just need to extract the list of numbers from our human, and the solution is ready (Figure 1-6)!

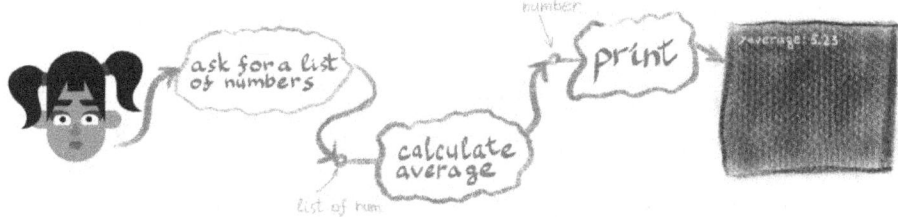

Figure 1-6. *Step 3 in resolving the problem*

What did we just do? We have broken down or *decomposed* our problem into smaller pieces, solutions to which we already know or can come up with easily enough, and combined them in the right order to get to the desired result. Just like with the mage above, we cannot give a string or a horse to our *calculate average* – it expects a *list of numbers*, so it *will not compose* with functions that produce something other than a list of numbers! Thus, we came up not only with "spells" – functions that we would need to solve our problem – but also with *types* that our functions should work with: a list of numbers, a number, an *effect* of putting something on screen. From the start, just by thinking about how to solve a problem, we "invented" two most important concepts in programming, computer science, and mathematics itself: **functions** and **types**.

CHAPTER 1   WIZARDS, TYPES, AND FUNCTIONS

We'll start looking at them more closely later in this chapter. For now, here's a bit of real magic: as it turns out, Haskell code maps virtually one to one to this approach! Look at the first line of code below (don't mind the rest for now):

```
main = ask >> getLine >>= toList >>= calcAverage
>>= printResult
      where calcAverage l = pure $ (fromIntegral $ sum l) /
      (lengthF l)
            printResult n = putStrLn $ "Average of your list
            is: " ++ (show n)
            ask = putStrLn "Enter list of numbers, e.g.
            [1,3,4,5]:"
            lengthF = fromIntegral . length
            toList :: String -> IO [Int] = pure . read
```

The first line is our sequence of "spells," or, really, functions, that start with a human sitting at the keyboard and get us to the line on the screen with calculated average shown. First, we ask a human to enter the list of numbers with `ask` and pass whatever was entered to the next function – `toList` – to convert the string she enters to an actual `List`. The resulting list is given to the `calcAverage` function that in turn passes its result to `printResult`. Exactly what we came up with, reasoning about the problem in the passage above.

We started with the desired end result, worked backward by decomposing the big problem into smaller pieces, and *connected* or *composed* them so that the output type of one function fits to the input types of the next. This is the essence of *every* functional program. Lines of code after `where` further detail how the "spells" we are chaining should behave – don't concern yourself with them yet. What's important is to get a feel of the process:

- What is our desired result?
- What is our input?

- Work backward from the result to the input, decomposing the problem into smaller pieces that we know how to solve, *by thinking about how to convert between types along the way.*

- *Compose* smaller functions, each of which solves a smaller problem piece, together into a bigger function, making sure the types fit.

Question: Why are we working backward from the desired result instead of forward from the input? For elementary tasks, such as the one above or turning a cockroach into a Ferrari, it doesn't really matter. When you start encountering more complicated real-world problems, it will be much easier to start with the type of the desired result – just like it is easier to solve a maze starting from an exit!

Practice this approach on some other problems you would like to solve. We will use it on quite a bit of the real-world problems further along.

Let's Fall in Love with Types!

To become a powerful Haskell mage, you need to get to know and understand **types** intimately: their birth, their daily lives, and their relations with each other. Types come before functions and functions are meaningless without types. So what are types? Intuitively, a type is just a bunch of objects (or subjects) sharing some common characteristic(s) thrown in a pot together. You could say that the type of your Ferrari Spider is *Car*, as is the type of that Toyota Prius. The type of both a pig and a horse is *Animal*. The type of 1, 114, 27428 is *Natural Number*. The type of the letters "a," "C," "e," "z" is *Char* (that's an actual type for characters in Haskell). The last two are ubiquitous in programming, so let's look at them more closely.

CHAPTER 1   WIZARDS, TYPES, AND FUNCTIONS

Let's not forget that we are creative wizards, so we will be building a whole new world from scratch. We will call this world **category "Hask"** (for reasons that will become apparent much later), or simply **Hask** from now on, and it will be inhabited by **types**; their members, or *values*; **functions**; and some other interesting characters. As we create this world, we will gradually become more and more powerful and eventually will be able to write efficient, elegant, and robust Haskell programs for any practical task.

A Char and an Int Walk into a Bar and Make a Function

Natural numbers are the most intuitive math concept that everyone is familiar with – that is what you use when counting the times you read confusing monad tutorial articles. Natural numbers start with nothing, zero, and go on forever. Once you subtract the number of times you read confusing monad tutorials (at least 37) from the number of times it made sense (0), you realize the need for negative numbers, which together with naturals give you another *infinite* type – integers. Various "ints" that you might be used to from the C-land, as well as the type **Int** in Haskell, are the teeny-tiny subset of real integers, limited by 2^31^-1 or 2^63^-1 on the 64-bit systems. What is 9,223,372,036,854,775,807 compared with infinity? Same as nothing, but, fortunately, it is enough for most practical purposes, including counting all the money in the world if you want to write a new automated banking system at some point.

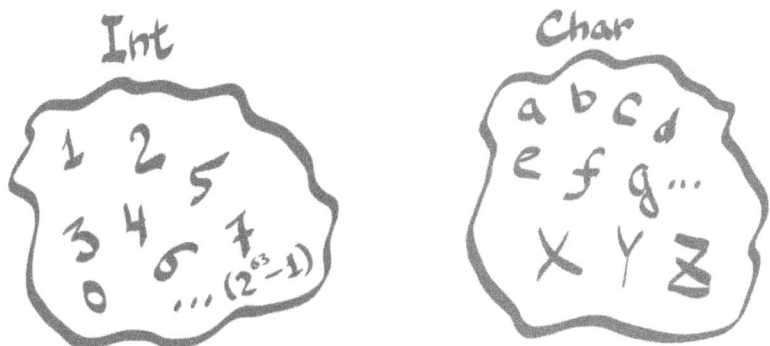

Figure 1-7. *Introducing Int and Char types*

Char is our second foundational type – a pot of all characters we use when writing stuff in different languages, including mathematics or Egyptian pictography if we want to get fancy. It is also a finite type and quite a bit smaller than **Int** at that.

By themselves, these types are quite boring (see Figure 1-7), and we can't do much with them – it's just lonely numbers and characters hanging out all by themselves. They are sort of like prisoners in solitary confinement with no way to interact with others. Let's break them out and set them free to roam the world and talk to each other! To do this, we are going to need **functions**.

Note What is called "functions" in the imperative world, be it JavaScript or C family, has very little in common with mathematical functions we are interested in. A JavaScript "function" better be called a "procedure" – as they used to be called in ancient times, before Facebook, when people still cared about being thorough, in languages such as Fortran. In a procedure, you tell a computer what to do – add some numbers, draw a circle, divide by zero – spaghetti of messy code thrown together. We want to be tidy, elegant, and comprehensible, so we will use real mathematical functions.

CHAPTER 1   WIZARDS, TYPES, AND FUNCTIONS

Let's line up our characters from the **Char** type as they do in an English alphabet. Then we can ask each letter, "Which number are you?", like in Figure 1-8.

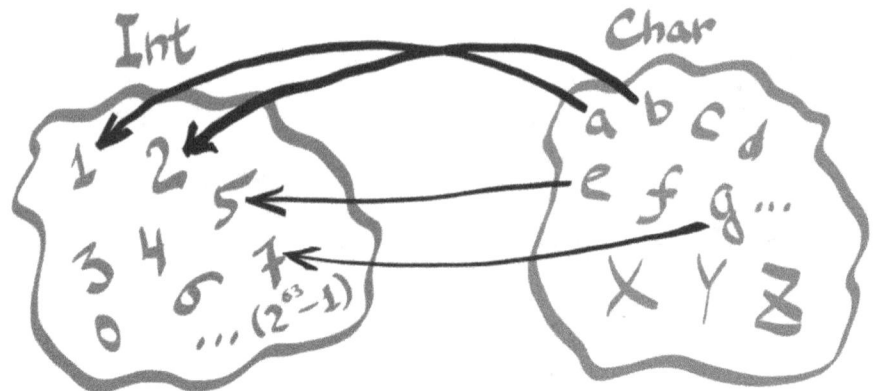

Figure 1-8. *Introducing functions*

We have just constructed our first function! What it does is match *each* letter with *a* number. Now, this is very important: it matches *all* letters with *exactly one* number. We could match just *some* letters to certain numbers – then our function would be called a *partial* function since it would be defined not for all members of our type. A function can map *several* members of a type to one member of another type, e.g., we could define a function that matches "a" and "b" to 1, "c" and "d" to 2, etc. – it would be a perfectly fine, even if a bit weird, function. However, what we *cannot* do is define a function that matches a member of a type to *several* members of another type (see Figure 1-9).

Figure 1-9. *Functions and not functions*

On the other hand, an imperative *procedure*, for instance, a JavaScript "function," can take an "a" and return 1 on one day, "32" on another, or crash if the temperature in Alaska drops below zero. Our **function**, as any other real mathematical function, *always* gives you the same result for the same arguments, no matter what.

You can trust the word of a function. You cannot trust anything from a procedure.

Our function (let's call it whichNumberIsChar) takes a **Char** and gives an **Int** as a result. In Haskell, we write: whichNumberIsChar :: Char -> Int. You can try it live in your browser here https://repl.it/languages/haskell or by installing Haskell environment from Haskell.org and starting your REPL locally. Just type (cheating a bit since we are using another, much more powerful *polymorphic* Haskell function to define ours) let whichNumberIsChar = fromEnum :: Char -> Int. Then you can type whichNumberIsChar 'Z' (or some other character) and see what happens.

Our world is still very boring. We have two basic types living in it, and we have one function between them, but we can't even add two numbers together! That's not enough to solve a simple *calculate an average for some numbers* problem, let alone do something more interesting. However, it turns out that **Int**, **Char**, and some imagination can get us *much* further than may appear. Let's make some more magic!

CHAPTER 1   WIZARDS, TYPES, AND FUNCTIONS

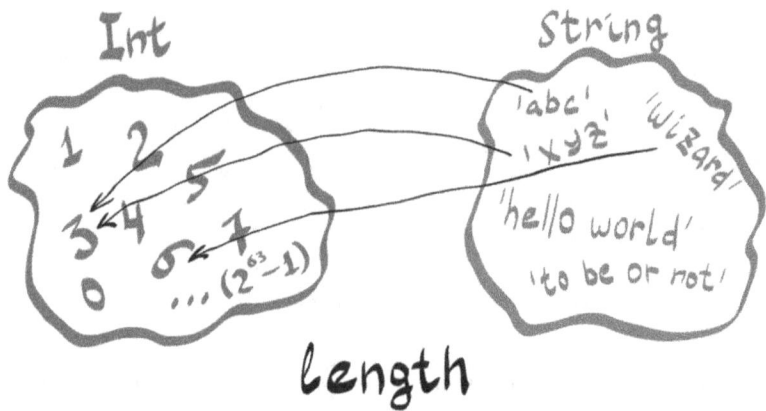

Figure 1-10. *More complex types and functions*

How FunTypical!

First, we are going to need strings. What is a string? It's basically just a bunch of characters put together in order. Like "abc," "xyz," or "Hello World." This hints at how we can introduce the type **String**: it's a set of all possible combinations of characters from **Char**. Easy, right? We can also introduce another obvious useful function right away, between **String** and **Int**, shown in Figure 1-10, which gives us the length of a string: length :: String -> Int. Try it out in REPL by running ghci in your terminal, and you should see something like

```
GHCi, version 9.4.8: https://www.haskell.org/ghc/   :? for help
ghci> length "abc"
3
ghci> length "hello world"
11
ghci> :t length
length :: Foldable t => t a -> Int
```

Now, the last command – `:t length` – is interesting. As you can guess it prints out the type of any object in Haskell – extremely useful information. As you can see from the result, our `length` function has a type that is much more generic than what we have assumed – instead of `String -> Int` it has a type `Foldable t => t a -> Int`. We will learn what exactly this means soon enough. For now it suffices to know that it indeed works on strings.

How about "some numbers" type? We can use the same logic as above and introduce the type **ListInt** – a *set of all possible combinations* of numbers from **Int**, such as [1, 2, 3] or [10, 500, 42, -18]. What if we treat each string as a sentence and want to define a type **Book** that would contain a bunch of sentences together? Again, we can define it as a *set[1] of all possible combinations* of strings from **String**.

Do you notice a pattern? We take a type, apply some manipulation we describe as *"a set of all possible combinations"* to members of this type, and get *a new type as a result*. Doesn't it look just like a function?! (*See Figure 1-11.*) We just "turned" one type into a different type, just like the function length turns a string into an int.

---

[1] Our manipulation procedure is not strictly creating "sets" in a mathematical sense of the word, since in a real set each object can only be present once. In our new "list" types, values may repeat – [1,1,1,1] is a perfectly valid list of ints in Haskell even though it's not technically a set.

CHAPTER 1 WIZARDS, TYPES, AND FUNCTIONS

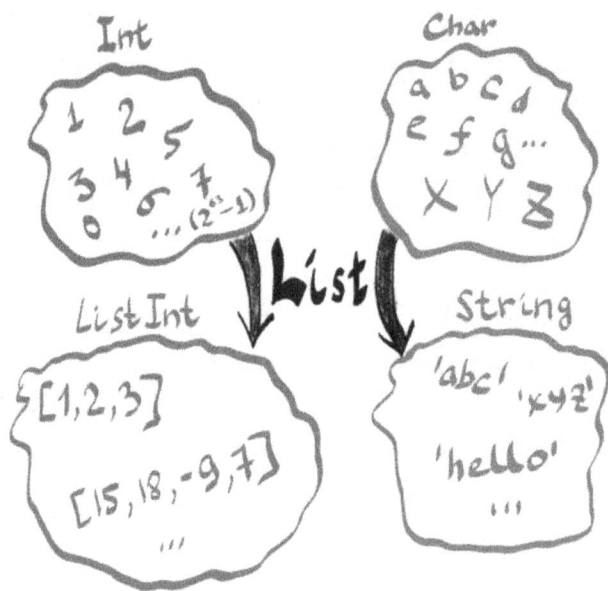

Figure 1-11. *Constructing new types from existing types*

Let's call this manipulation `List` – just like in the picture. As we have defined functions `whichNumberIsChar :: Char -> Int` and `length :: String -> Int`, `List` is also indeed a function, one that takes a **type** and creates a *different* **type**! In Haskell, we write `List :: * -> *` (`*` is a synonym for concrete **Type** in Haskell, so in fact this signature means `List :: Type -> Type`). They are sometimes more formally called *type-level functions,* and they are the core of the beauty, elegance, and efficiency of Haskell. We will be doing *a lot* of type creating down the line!

Info Now, since our String and ListInt were created in a very similar way, can we apply our length function to it? Try it out in REPL: length [1,2,3], length [], give it some other lists of numbers. Did it work?

14

CHAPTER 1   WIZARDS, TYPES, AND FUNCTIONS

Curry and Recurse!

We saw that a function takes one argument of one type and maps it to exactly one value of a different or the same type. A couple of questions should arise at this point:

- What do we do with functions of several arguments that exist even in math - like f(x,y) = x + y - not just C-land programming languages?

- How do we actually define functions without resorting to writing explicit tables that map arguments to results, something like "1 + 1 = 2, 0 + 1 = 1, 1 + 0 = 1 ..." for all numbers? You could certainly do that, but it's a bit unpractical, where by "unpractical" I mean totally nuts.

There are a couple of ways to tackle the first one. Let's say we want a function that takes a **Char** and an **Int**, converts a character to a number with the help of our `whichNumberIsChar` function, and calculates the sum of the two, something like `anotherWeirdFunction c i = (whichNumberIsChar c) + i` - but we can't do this since the functions we defined take only one argument! One way to work around it would be to somehow combine our **Char** and an **Int** into a new type of pairs (**Int, Char**) - similar to what we have done with **List** above - and write

```
anotherWeirdFunction :: (Char, Int) -> Int
anotherWeirdFunction (c, i) = (whichNumberIsChar c) + i
```

Now we are ok - our function takes just one argument. This is a perfectly valid way to do it, and we will see how to construct pairs from other types in the next chapter. However, it turns out it is much more powerful to use a different approach. First, one very important point:

Functions are first-class citizens in functional programming (that's kind of why it's called *functional* programming) - which means you can pass a function to another function as an argument and return a different

15

function as a result. This also means that we have types that are inhabited by *functions* – e.g., type of functions `String -> Int` would contain all possible functions that take a string and return an int value.

Let's say you want to take a **ListInt** we defined above and multiply all its elements by some number n. In the imperative world, you would write a function that takes a list and a number n, goes through some cycle, and multiplies it. An okay approach, but what if later on you'd need to *add* a number n to all list elements? You'd need to write a different cycle that now adds numbers. Wouldn't it be great if we could just pass a function that we want to apply to all list elements, whatever it is? In Haskell, you can, by simply writing

```
map (*2) [1,2,3]
=> [2,4,6]
map (+2) [0,8,18]
=> [2, 10, 20]
```

We pass a function that adds 2 – (+2) – or multiplies by 2, (*2), and a **ListInt** to a function called map (we will define it in the next chapter) and get another list as a result. Much more concise and powerful than writing a lot of boilerplate imperative cycles and also hints at how we can define functions of multiple arguments by looking at some further examples:

- (+) x y is a function of one argument that takes an Int and returns a *new function* that takes an Int and returns an Int. So (+2) is a function that adds 2 to any number, (+18) adds 18, etc. – and we create these functions on the fly by passing only one argument to (+) first. So the type signature of (+) is (+) :: Int -> Int -> Int, which you can read as (+) :: Int -> (Int -> Int), i.e., exactly what we just wrote – it takes an Int and returns a function that converts Int to Int!

- Our anotherWeirdFunction is defined similarly: it's a function of one argument that takes a Char and returns a function of one argument that takes an Int and returns an Int:

```
anotherWeirdFunction :: Char -> Int -> Int
anotherWeirdFunction c i = (whichNumberIsChar c) + i
```

So, in general, a function of many arguments can be defined as a function that takes one argument and returns a function that takes one argument and returns a function ... This technique is called *currying*, because that's how you make chicken curry. No, not really. It was invented by Haskell Curry, in whose name both the currying *and* the Haskell language itself were named.

Currying is extremely useful, because it makes the math much simpler and allows you to create different functions on the fly and pass them around as you see fit.

Now, let's turn to the second question – **how to construct functions efficiently**.

It is in fact possible to start with an empty set and define all of Haskell – first, you construct natural numbers starting with an empty set as we showed in an aside above; then you can define an operation of addition on these numbers, invent integers once you realize you need negative numbers, and then define multiplication via addition; etc. It is a long and winding road.

We won't be doing that (although it's a pretty cool exercise to do on your own) – since existing computer architectures provide a bunch of operations, such as addition, multiplication, and memory manipulation, for us much more efficiently than if we were to implement them from scratch by translating pure functional lambda calculus logic. Haskell additionally defines a bunch of so-called *primops*, or primitive operations,

CHAPTER 1   WIZARDS, TYPES, AND FUNCTIONS

on top of those and then a bunch of useful functions in its standard library called **Prelude** – so you can obviously use all of them when writing your functions and do things like

```
square x = x * x
polynom2 x y = square x + 2 * y * x + y
```

Having said that, a very important concept in functional programming that we need to understand is *recursion*. It allows you to write concise code for some pretty complex concepts. Let's start with the "Hello World" of functional programming – factorial function. Factorial of a natural number n is defined simply as a product of all numbers from 1 to n, and in Haskell you can simply write product [1..n] – but it's too easy, so let's create our factorial function recursively:

```
fact :: Int -> Int
fact n = if n == 0 then 1 else n * fact (n - 1)
```

What's happening in recursion is that a function calls itself under certain conditions – in the case of fact, when n > 0. Try executing this function on a piece of paper with different arguments. For example, if we "call" fact 1, here's what's going to happen:

```
[1] fact 1
[2] if 1 == 0 then 1 else 1 * fact (1 - 1)
[3] if False then 1 else 1 * fact 0
[4] 1 * fact 0
[5] 1 * (if 0 == 0 then 1 else 0 * fact (0 - 1))
[6] 1 * (if True then 1 else ...)
[7] 1 * 1
[8] 1
```

Phew. Now try expanding fact 10 on paper when you have half an hour to waste. There's nothing complex about recursion – you (or a computer) simply substitute(s) a function call to the function's body as normal, and the main thing to remember is that your recursion needs to terminate at some point. That's why you need boundary cases, such as fact 0 = 1 in the case of factorial, but it can be completely different conditions in other functions.

EXERCISES

1) Our fact function as defined above has a problem. Can you find and fix it?
2) Define a multiplication function via addition recursively.
3) A very interesting case of recursion is the Ackermann function: https://www.wikiwand.com/en/Ackermann_function. Write it in Haskell but don't try calling it with more than ackermann 4 1 — even this case may take you some hours to compute. Practice expanding ackermann 1 1 on paper to understand internal recursion machinery.

Conclusion

In a very short time, starting with barely anything but an empty set, we have come up with the concepts of

- Types, inhabited by their members, also called *values*. We have introduced two very simple types: Int and Char.

- Functions, which map values of one type to values of another (or the same) type.

- Type functions, which create new types from other types. We have introduced a couple of useful complex types: String and ListInt.

- Composition, an essence of a functional program (and, incidentally, Category Theory, but we won't be going there for quite some time yet).
- Ways to construct functions via currying and recursion.

This is pretty much everything we need to gradually build our magical **Hask** world!

What is currently missing is details on *how* exactly we can define and use functions such as `length` that gives us the length of a **String** (or, as we have seen above, any other **Type** constructed with the help of our `List` type function!) and `List` that creates new types. We can't just tell a computer, *"Give me a set of all possible combinations"* – we need to be slightly more specific.

In the next chapter, we will continue following the adventures of our mage and will focus on card magic while constructing different types and intimately understanding their internal structure – which is an absolute requirement for designing efficient functions.

Let's roll up our sleeves and get to it!

CHAPTER 2

Type Construction

You got to "Nobu" right on time, barely past seven. As you were approaching the hostess, a hurricane of a girl ran into you – "Thank goodness you got here on time, Curry!" *Curry, that must be my name? But why? I don't even like curry?! Must be a different reason ...* – grabbed your hand, and dragged you past the busy tables somewhere deep in the restaurant, through the kitchen, out the back door, down the stairs, and into a dimly lit room. In the middle of the room stood a card table – green cloth, some barely visible inscription that spelled *"Monoid"* in golden letters on it – surrounded by six massive chairs that looked really old. The girl abruptly stopped, turned around, looked you right in the eye – hers were green – and asked almost irritably, but with a pinch of hope in her voice:

- Well? Are you ready?

- Ahem ... Ready for what?! Can you explain what's going on?

- Did you hit your head or something? And where were you last week? I called like 27 times! And why do you smell like manure? Did you suddenly decide to visit a farm?! *Oh, the exploding horse, right ...* Ah, no time for this!

She dragged you to the table and pushed you into one of the chairs – *she was strangely strong despite being so miniature* – so that you had no choice but to sit. BOOM! Suddenly, a small firework exploded in your eyes, and you felt *more magic* coming back to you.

Out of thin air, you create a new deck of cards with intricate engravings on their backs, nothing short of a work of art. You feel them – they are cold and pleasant to touch – but then rest your hands on the table, close your eyes, and ... It's as if the cards became alive and burst in the air above the table, shuffling, flying around, dancing. You open your eyes, enjoying your newly mastered power over cards, and barely nod your head, and the deck, obeying your wishes, forms into a fragile house of cards right beside the *Monoid* sign.

You turn to look at your mysterious companion. She is smiling and there are twinkles in her eyes.

House of Cards with a Little Help from Algebra

Looks like our mage is making friends and remembering new spells, this time without risking any animal's well-being. Let's follow in his footsteps and learn some card magic!

First, download this chapter's code from GitHub and try dealing some cards. You'll see some beautiful (well, not as beautiful as the ones Curry created, but still ...) cards being dealt – 9♥ 10♥ J♥ Q♥ K♥ A♥ – with barely more than 20 lines of code. Let's understand what's going on here step by step.

To start doing card magic – or card programming for that matter – we also need to create a deck of cards out of thin air. Here's one way of doing it in Haskell:

```
-- Simple algebraic data types for card values and suites
-- First, two Sum Types for card suites and card values

data CardSuite = Clubs | Spades | Diamonds | Hearts deriving (Eq, Enum)
data CardValue = Two   | Three | Four | Five  | Six  | Seven
 | Eight | Nine | Ten   | Jack | Queen | King | Ace deriving (Eq, Enum)

-- Now, a Product Type for representing our cards:
-- combining CardValue and CardSuite
data Card = Card CardValue CardSuite deriving(Eq)
```

We have created three new types above. **CardSuite** is inhabited by four values, Clubs, Spades, Diamonds, and Hearts – just like **Int** is inhabited by 2^{64} values – that correspond to playing card suites. **CardValue** is inhabited by 13 values that represent card face values. All of it is pretty self-explanatory. Type **Card** represents a specific card and is a little bit more interesting – here we are combining our two existing types into one new type. Don't mind the `deriving` statements for now – we'll explain them down the road. Suffice it to say that here Haskell automatically makes sure we can *compare* values of these types (Eq) and that we can convert them to numbers (Enum).

Open REPL linked above again, run it (if you don't run it first, you'll get an error as the types won't be loaded into the interpreter), and then play with these types on the right:

```
> let val = Jack
> val
=> J
```

CHAPTER 2   TYPE CONSTRUCTION

```
> let suite = Spades
> suite
=> ♠
> let card = Card Ace Hearts
> card
=> A♥
> let card1 = Card val suite
> card1
=> J♠
```

Here we bound val to **CardValue** Jack and suite to **CardSuite** Spades and created two cards – Ace of Hearts directly from values and Jack of Spades from our bound variables – both approaches are perfectly valid. Create some other cards in REPL to practice.

Looking at the types' definition above, the syntax for type creation should become apparent: to create a new type in Haskell, we write data <Name of Type> = <Name of Constructor1> <Type of value 11> ... <Type of value 1n> | <Name of Constructor2> <Type of value 21> ... <Type of value 2m> | .... You can read symbol "|" as "or" – it is used for combining different types together into a new *sum type*, values of which can be taken from a bunch of different types we are combining, but every value is taken from only one of them. Types like data Card = Card CardValue CardSuite are called *product types*, and every value of such type contains values of *all* of the types we are combining.

EXERCISE

Now that you know how to define new types, think about how you would add a Joker card to your Card type? Amend the data Card = ... definition accordingly.

Important The type name and constructor name do not need to be the same; you could have defined our Cards type, for instance, as data Cards = Card CardValue CardSuite. However, when you define a type that has only one constructor, it often makes sense to use the same name as we did above.

Algebraic Data Types

Sum and product types together are called **algebraic data types**, or **ADTs** in short (*which may be confused with abstract data types that are quite a different matter, but we will call algebraic types ADTs*), and they form a foundation of any modern type system.

Understanding ADTs is extremely important in Haskell programming, so let's spend more time discussing them and looking at different ways to define them. Here are several more examples.

Type **Person** contains their name and age: data Person = Person String Int. We are combining a **String** to represent a name and an **Int** to represent an age, and then we can create people like so: Person "Curry" 37 or Person "Jane" 21. This is a product type.

What if we wanted to represent built-in JavaScript types in Haskell – String, Number, Date, Object, etc.? Here is one way of doing it (it does not allow us to represent everything; we will look at the full JavaScript value type definition once we learn recursion):

```
data JSValues = JString String | JDate Int | JNumber Double | Undefined
```

CHAPTER 2　TYPE CONSTRUCTION

Now, this is interesting – we have combined three types that "store" a String, an Int, and a Double inside a constructor and added a simple value "Undefined" as the fourth possible value. It means that values like `JString "hello world"`, `JDate 12345225` (which is quite a long time ago), `JNumber 3.141`, and `Undefined` are all valid values of the type **JSValues**. This is a sum type (see Figure 2-1 for representation of the types discussed).

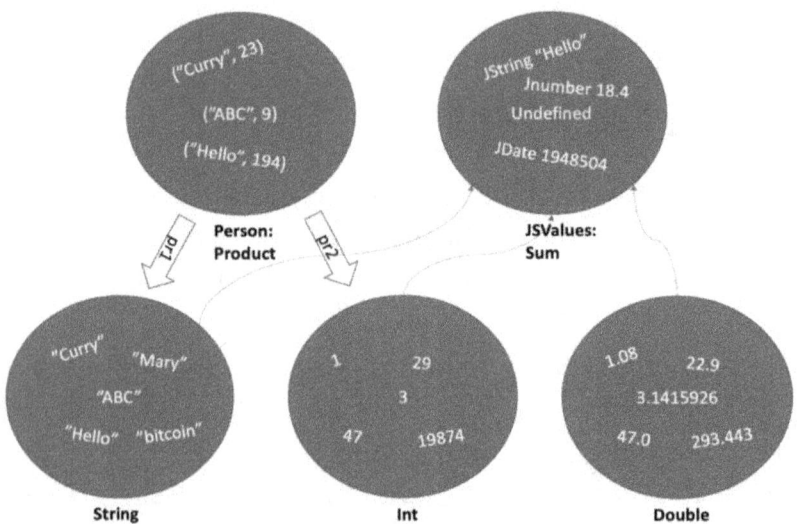

Figure 2-1. Sum and product types

By now, you should get a pretty good feel on what sum and product types are. When you create a product type, you combine values of several different types in *one* value, represented as a *tuple* (`x_1:t_1, x_2:t_2, ...`). When you create a sum type, each value of your type can be from a different type, but only one.

Practice creating types for different tasks you would be interested in solving.

Records

What if you wanted to extend the type **Person** to include more information about a person, which would be much more useful in a real-world situation? For example, place of birth, parents, first and last names, education, address, etc. If you had to write something like

```
data Person = Person String String Int Date Address Person Person
jane = Person "Jane" "Doe" 25 (Date 12 10 1992) addr
father mother
```

it would be extremely confusing to use - you'd need to remember the order of fields, you'd confuse fields with the same types, etc. Thankfully, Haskell supports *records* - basically, a *product type* with named fields. We can define our Person type properly like this:

```
data Person = Person
  {
    firstName :: String
  , lastName  :: String
  , age       :: Int
  , dob       :: Date
  , address   :: Address
  , father    :: Person
  , mother    :: Person
  }
-- you can still initialize a record just by applying a
Constructor ('Person') to a bunch of values in order
jane = Person "Jane" "Doe" 25 (Date 12 10 1992) addr mother father
-- ... or, you can access the record's fields separately by
their respective names
janesSister = jane { firstName = "Anne", age = 17, dob = Date 1
1 2000 }
```

CHAPTER 2 TYPE CONSTRUCTION

PERSON EXERCISE

The type **Person** defined above has a problem – the "father" and "mother" fields create a recursion, which we have no way to terminate. Can you think how to solve this problem using sum types? If not, read the next section about **Maybe,** and it should give you an idea.

Heads Up   Records are in fact one of the biggest problems in Haskell right now due to somewhat inconvenient syntax. There are solutions in the form of a couple of libraries, such as lenses, and there's active discussion on how to improve record handling in the future. But despite those syntactic inconveniences, they are fully functional records that are actively used in real-world programs.

Before we move on to the other ways of constructing new types, here is yet another example, which illustrates how product and sum types work together. Let's say we want to model different two-dimensional geometric shapes. We can define the following product types:

```
data Point = P Double Double -- x and y
data Circle = C Double Double Double -- x and y for the center
coordinates, radius
data Rectangle = R Double Double Double Double -- x and y for
the top left corner, width and height
```

Now, let's say we want a function that calculates the area of our shapes. We can start like this:

```
area :: Rectangle -> Double
area (R x y w h) = w * h
```

```
-- since we are not using x and y values from our Rectangle to
calculate the area, we can omit them in
-- our pattern match like so:
area (R _ _ w h) = w * h
```

This works fine. But we need to calculate the area of circles as well and potentially other shapes. How do we do it? If you try to define another function `area :: Circle -> Double`, the typechecker will complain because you already have a function named `area` that has a different type. So what do we do? Create a differently named function for every type? Now, that would be ugly!

SHAPES EXERCISE

Think about how you would amend our types' definitions to make the function `area` polymorphic. Don't read further before you do!

There are several approaches to solving this. (*This is actually one of the problems with Haskell – you can solve one task in many different ways, and there's just so much power and flexibility that sometimes it is getting overwhelming, and not just for beginners. We will see many examples of this down the road.*) One of them is using **typeclasses** or even **type families** (probably an overkill in this case) – we'll get there soon. But now we will simply use a sum type. We can combine three of our shapes (product types) into one sum type. There are actually two ways to do it! The first one is if we want to keep Rectangle, Circle, and Point as separate concrete types in our system. Then we will use the definition above and will combine them into one type like so:

```
data Shapes = PointShape Point | CircleShape Circle |
RectangleShape Rectangle
```

CHAPTER 2   TYPE CONSTRUCTION

Then, our area function can be defined as follows with the help of pattern matching:

```
-- notice the type signature change:
area :: Shapes -> Double
-- we don't care about the point coordinates, area is always 0
area (PointShape _) = 0
-- notice the nested pattern match below:
area (CircleShape (C _ _ r)) = r * r * pi
area (RectangleShape (R _ _ w h)) = w * h
```

Now everything will work like a charm. When you write somewhere in your code area myShape, it will check what exact shape myShape is – Point, Circle, or Rectangle – and will choose the respective formula.

An alternative way to achieve the same result is to define the Shapes type directly as a sum type like so:

```
data Shapes = Point Double Double | Circle Double Double Double
| Rectangle Double Double Double Double
```

I leave the area function definition for this case as an exercise. The difference from the previous approach is that in this case you will not have types Point, Circle, and Rectangle in your type system – they are simply constructor functions now. So trying to define a function from Rectangle to Double will be a typechecker error.

Using sum and product types, you can express a lot of real-world modeling already, but why stop there? We want to be even more efficient and polymorphic! Read on.

Type Functions Are Functions, Maybe?

Let's spend some time reviewing a statement we made in the last chapter: *type and data constructors are functions.* This is extremely important to understand and internalize, so that you do not become confused when writing real-world programs that use monad transformer stacks and similar concepts that may look overwhelming, but are in fact quite simple. **You need to be able to easily read type signatures and understand their structure at first glance.** This is *the* key to mastering Haskell.

Let's look at a couple of types, renaming type names now to distinguish between types and constructors: data Cards = Card CardValue CardSuite and data People = Person String Int. The part on the left is the name of the *type*, and the part on the right is the name of the *data constructor* and types of the values that this constructor contains. So, in reality, *Card* is a *function* that should be read in Pascal-ish pseudocode as function Card (x:CardValue, y:CardSuite) : Cards, i.e., it's a (constructor) function from two variables, one of type CardValue and another of type CardSuite, that returns a value of type Cards! Same with *Person*, which turns into function Person (x:String, y:Int) : People.

In fact, using so-called GADT (generalized algebraic data type) Haskell notation, this becomes even more apparent – below is the solution to the task of extending **Card** type with the *Joker* card written first with the familiar syntax and then using GADT:

```
data Cards = Card CardValue CardSuite | Joker

-- same using GADT
data Cards where
  Card  :: CardValue -> CardSuite -> Cards
  Joker :: Cards
```

Looking at the second variant, the point about *Card* and *Joker* being functions becomes much clearer. In effect, it tells us that we are defining type **Cards**, which contains two constructors: **Card**, which takes two arguments of types CardValue and CardSuite and returns a value of type **Cards**, and **Joker**, which takes no arguments (or, strictly, an empty type – ()) and is a value of type **Cards**.

CHAPTER 2   TYPE CONSTRUCTION

We can take it one step further and take the analogy of *product* and *sum* types to its fullest – we can treat the type **Cards** as a *type function* that is a *sum* of two *product* types: typeFunction Cards = Card (x : CardValue, y : CardSuite) + Joker()! Of course "+" here has nothing in common with arithmetic sum but has the same meaning as "|" in actual Haskell syntax.

So **Card** is a constructor that creates a *value of type* **Cards** from two other values, and **Cards**, even though we just called it a type function, is in fact a "type constant." It is a concrete type – you cannot apply it to another type and construct a new one. Unlike type function List that we hinted at in the last chapter, which actually creates new types from other types and as such is much more interesting and important for us. We'll get to it in the next section, and we'll need to understand it to follow along our card dealing program further.

For now, let's start with the simplest real (non-constant) type function, one that constructs new types from other types: **Maybe**, which is used ubiquitously in Haskell:

```
data Maybe a = Just a | Nothing
```

```
-- in GADT syntax:
data Maybe a where
   Just    :: a -> Maybe a
   Nothing :: Maybe a
```

CHAPTER 2 TYPE CONSTRUCTION

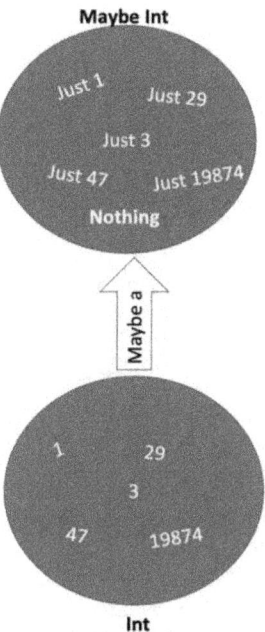

The notation above is very often a source of confusion for beginners and even advanced Haskell students – **we are using variable "a" on both sides of the type definition, but they mean very different things on the left and on the right!** In fact, this confusion should be cleared if you read this sort of definition in the spirit of what we've described above: typeFunction Maybe (a : Type) = Just (x : a) + Nothing ().

Please dwell on it until you fully understand the issue – it will save you lots of hours of frustration with more complex types down the road. We are in effect saying here: *Maybe is a type function that takes a **type** variable "a" and creates a **new type** from it ("Maybe a") with the help of two constructor functions – "Just," which takes a "regular" variable of **type a** and creates a **value** of type "Maybe a," and "Nothing," a function of zero arguments that is a value of type "Maybe a."* So "a" is a variable of type "Type" – it can take a value of any Type we defined – on the left, and it is used in the type signature for constructor function "Just" variable on the right.

33

CHAPTER 2   TYPE CONSTRUCTION

Take a breath and let it sink in.

Why is the type (which really is a type function) **Maybe a** useful? It's the easiest way to handle failure in your functions. For example, you could define a function `divide` like so:

```
divide :: Double -> Double -> Maybe Double
divide x 0 = Nothing
divide x y = Just (x / y)
```

This function would return `Nothing` if you try to divide by zero, instead of crashing or throwing an exception, and `Just <result of division>` if you divide by non-zero. This is a trivial example, but **Maybe** is of course very useful in thousands of other more complex situations – basically, anywhere you need to handle the possibility of failure, think about **Maybe** (there are other ways to handle failure and errors in Haskell as we shall see further on).

Note also function definition syntax – a very powerful and concise *pattern matching* approach. Under the hood, this definition translates into something like

```
divide x y = case y of
    0 -> Nothing
    _ -> Just (x/y)
```

...but in Haskell you can avoid writing complex nested `case` statements and simply repeat your function definition with different variable patterns, which makes the code much more readable and concise.

So what if we also have a function `square x = x * x` that works on Doubles and need to apply it to the result of our safe `divide` function? We would need to write something like

```
divideSquare x y = let r = divide x y in
                   if r == Nothing then 0
                   else <somehow extract the value from under
                   Just and apply square to it>
```

To do the "somehow extract" part in Haskell, we again use pattern matching, so the function above would be written as

```
-- function returning Double
divideSquare x y = let r = divide x y in
                   case r of Nothing -> 0
                             Just z  -> square z

-- or, better, function returning Maybe Double as returning 0
-- as a square of result of divide by zero makes no sense
divideSquare x y = let r = divide x y in
                   case r of Nothing -> Nothing
                             Just z  -> Just (square z)
```

We pattern-match the result of divide and return Nothing if it's Nothing or Just square of the result if it's Just something. This makes sense, and we just found a way to *incorporate* functions that work between Doubles into functions that also work with Maybe Doubles. But come on, is it worth such a complicated verbose syntax?! Isn't it better to just catch an exception once somewhere in the code and allow for division by zero?

Yes, it is worth it and, no, it's not better to use exceptions, because you can rewrite the above as simply divideSquare x y = square <$> divide x y, or looking at all the function definitions together:

```
square :: Double -> Double -> Double
square x = x * x

-- safe divide function
divide :: Double -> Double -> Maybe Double
```

CHAPTER 2   TYPE CONSTRUCTION

```
divide x 0 = Nothing
divide x y = Just (x / y)

-- our function combining safe divide and
divideSquare :: Double -> Double -> Maybe Double
divideSquare x y = square <$> divide x y

-- compare with the same without Maybe:
divide' x y = x / y -- will produce an error if y is 0

divideSquare' :: Double -> Double -> Double
divideSquare' x y = square $ divide' x y
```

Wow. Isn't it great? We are using practically the same syntax for both unsafe functions between Doubles and safe functions between Maybe Doubles, and what's more, we can use *the same approach for a lot of other different type functions, not just Maybe a.*

Here, operator **$** is just function application with lowest precedence, so allows to avoid brackets – without it you'd write square (divide' x y) – and **<$>** is another operator that *also* applies functions defined for concrete types, but it does so for values *built with a special class of type functions from these types, e.g., Maybe a or List a*, without the need to write nested case statements. It is a key part of something called a **Functor** in both Haskell and Category Theory, and we will look at it in more detail in the next chapter.

In the end, it's all about making function composition easy but safe – remember Chapter 1 and how to think about solving problems functionally? You break the problem down into smaller parts and then compose functions together, making sure input and output types "fit together." All the complicated scary terms you may have heard or read about in Haskell – Functors, Monads, Arrows, Kleisli Arrows, etc. – are all needed (mostly) to make composing different functions easy, safe, and concise, which, in turn, makes our task to decompose a problem in smaller pieces so much more easy and enjoyable. So don't be scared, it will all make perfect sense.

> **FIXING PERSON EXERCISE**
>
> Think about some problems you had solved using exceptions in imperative languages that you may solve more elegantly using Maybe. Modify the type "Person" so that you can store information about children, which may or may not exist, for a person.

Advanced: On Data Constructors and Types

It pays to understand how Haskell (GHC specifically) treats and stores values of different types. One of the powerful features of Haskell is that it is a statically typed language – which means there is *no* type information available at runtime! This leads to much better performance with at the same time guaranteed type safety, but is a source of confusion for many programmers coming from the imperative background.

Let's look at the type **Person** we defined above, changing the names a bit to avoid confusion:

```
data Person = PersonConstructor String Int
```

```
-- or, in a clearer GADT syntax:
data Person where
  PersonConstructor :: String -> Int -> Person
```

The name of the type is **Person,** and it has only one data constructor called **PersonConstructor** that takes an **Int** and a **String** and returns a "box" of them that has a type **Person**. Technically, it creates a tuple (n :: String, a :: Int) – so, for instance, when you write something like john = PersonConstructor "John" 42, you get a tuple ("John", 42) :: Person. The beauty of Haskell is that during the typechecking stage, the compiler has only one job – *make sure your functions compose*, i.e., input types to all the functions correspond to what the function expects, and the output type of whatever the function's body produces corresponds to what is expected in the type signature as well. Once this check is done, all the type information gets erased – it is not needed, as we have guaranteed and *proved it mathematically* that every function will receive correct values at runtime. This immensely increases performance. That's why they say "if it compiles, it works" about Haskell programs.

So, if you have a function showPerson :: Person -> String that is defined something like showPerson (PersonConstructor name age) = name ++ " is " ++ show age ++ " years old" and then you call it on person John from above, here is what happens:

```
st = showPerson john
-- gets expanded into:
showPerson (PersonConstructor "John" 42)
```

Under the hood it is simply a tuple ("John", 42), and according to the pattern match on the left side of the showPerson definition, we get name = "John" and age = 42. Then these values get substituted in the function body, and we get a resulting string. But nowhere is type **Person** used! The typechecker *guarantees* that the first value in a tuple received by showPerson will have type **String**, while the second will have **Int**, so it is impossible to have a runtime error!

In a general case, if you have a data constructor <Cons> T1 T2 ... TN that belongs to some type **MyType** and then you call it to create a value as <Cons> x1 x2 ... xn, Haskell stores it as a tuple (x1,x2, ..., xn) of type **MyType**. The type is used during the typechecking stage and then gets erased, while type safety is guaranteed (*proved*) by math.

For sum types, the situation is a bit trickier. Recall our type of **Shapes**:

```
data Shapes = Point Double Double | Circle Double Double Double

area :: Shapes -> Double
area (Point _ _) = 0
area (Circle _ _ r) = pi * r * r
```

It has in this case two data constructors – **Point** and **Circle**, which behave exactly as described above. Function "area" during compilation gets rewritten using case analysis, in pseudocode:

```
area (s :: Shape) = case <constructor of s is> of
   Point -> 0
   Circle -> <get 3rd element of the (x,y,r) tuple and
   substitute for r in pi * r * r>
```

This means that for sum types of more than one data constructor, Haskell needs a way to distinguish between data constructors with which this or that tuple has been created. So, even though information that s is a **Shapes** type is erased, we keep the constructor number *inside the type* together with our tuple. So that Point 4 0 would be represented something like (0, 4, 0) (in reality it's done differently – more efficiently – but this illustrates the concept well enough) where the first value of the tuple is the constructor tag. GHC uses a lot of very smart optimizations under the hood where this approach has minimal possible overhead at runtime as well.

List, Recursion on Types and Patterns

Let's go back to our card dealing magic spells. To (almost) fully understand it, there's only one extremely important piece missing: **List** type function and recursively defined types, which are extensively used in functional programming to implement various data structures. If you recall, in

CHAPTER 2   TYPE CONSTRUCTION

Chapter 1 we described **List** as some manipulation that takes all possible combinations of values of another type and creates a new type from it. Try to define the List type function yourself before reading further.

Here is how we can operationalize this definition:

```
data List a = Cell a (List a) | Nil

-- or, in GADT syntax:
data List a where
  Nil  :: List a
  Cell :: a -> List a -> List a
```

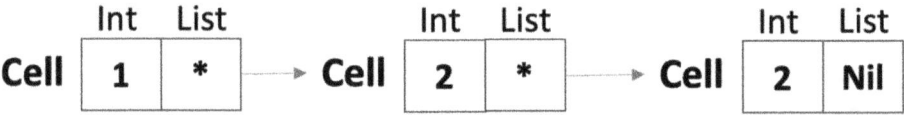

Figure 2-2. *List structure*

Remember how you need to read type definitions as type functions? In pseudocode – type function List (a : Type) = Cell (x:a, y: (List a)) + Nil() – **List a** is a type function from the variable a : Type that consists of two constructors: Nil, which is itself a value of type **List a**, and Cell, which takes a value of type a and a value of type **List a** (*this is where recursion comes into play!*) and returns another **List a**. Here Nil is the boundary case of our recursion, and Cell is a recursive function that constructs a list until it is terminated by the Nil value as the last element (see Figure 2-2).

LIST EXERCISE

Try constructing different lists, similar to how you "executed" the "fact n" function in the last chapter, from different types – make some lists of Ints, Strings, Chars, etc.

Let's construct a list from values 1, 2, and 3. You have a constructor function Cell that takes two values – a value of some concrete type, in this case an **Int**, and a **List** of **Int**. What is the simplest value of type **List**? Of course it's a Nil! So we can write Cell 3 Nil – and it will be a **List Int**! Now, we can pass this list to another Cell call, like this: Cell 2 (Cell 3 Nil) – and we have yet another **List Int**. Apply it one more time, and you get Cell 1 (Cell 2 (Cell 3 Nil)). Easy, right?

Basically, if you keep in mind that type functions and constructors *are*, well, functions, which follow all the same principles as "real" functions, this type building business should become very clear and easy for you.

TREE EXERCISE

Now, quick, grow a tree! Define a recursive tree type function – one that can create trees from any other types, similar to List.

We looked at the principle of how **List a** type function can be defined, but since in Haskell it is such a fundamental data structure, it is in fact defined with different constructor names and some syntactic sugar on top: Nil is in fact [] and Cell is defined as an operator :, so that our list of 1, 2, and 3 would be written as 1:2:3:[] or simply [1,2,3] – much more concise and convenient. You can write ["abc", "hello", "world"], a value of type **List String**; [1.0, 3.1415926, 2.71828], **List Double**; etc.

In the last chapter we introduced the type **String** as a set of all possible combinations of characters, or simply **List Char**. In fact, it is *exactly* how **String** is defined in Haskell – which is quite logical from the math point of view, but at the same time very inefficient in practice. It's kind of the same as if we defined our numbers as 0 being an empty set, 1 being a set containing an empty set, and so on and then defined operations of addition and multiplication on top of it – instead of using CPU-supported ints, floats, and operations on them. It would work, and might look very beautiful mathematically, but would be thousand times slower than asking a CPU to compute 2 + 2 via native operations.

In a similar fashion, when you are writing real-world programs that do any noticeable amount of string manipulation, you should try to use the type **Text** instead of **String**. It is represented much more efficiently internally, fully supports Unicode, and will make performance generally better. With lists, the fastest operation is appending an element *in the beginning* of the list – x:list. Inserting something in the middle or in the end is a pain that grows proportionally to the length of the list (in computer science we write O(n)).

To use **Text**, you should write

```
{-# LANGUAGE OverloadedStrings #-}

import Data.Text

s :: Text
s = "Hello World"
```

The first line here tells GHC to treat string literals as Text. Without it, you'd have to write a lot of annoying String-to-Text-and-back conversion statements. The second explicitly imports the **Text** data type for you.

Strings aside, lists are still very convenient for a lot of common tasks, as we will see down the line. Now that we know how to define functions and types, including recursive ones, we have a very powerful toolbox at our disposal already to start building something serious. Let's practice by defining some useful functions.

Length of a List

Remember this from Chapter 1? Now that you know how a list is built, you can easily write it:

```
length :: [a] -> Int
length [] = 0
length (x:xs) = 1 + length xs
```

Here we use recursion (length calls itself) and pattern matching: if a `length` is applied to an empty list (`[]`), it returns 0 (boundary case for the recursion!), and if it is applied to a list, we return 1 plus `length` applied to the remainder of the list. `(x:xs)` is pattern-matching a specific constructor – x here is simply the first element of the list, and xs is the rest of the list. It might make more sense if we wrote it via `Cell` constructor: `length (Cell x xs) = 1 + length xs`. If you recall, `Cell` (or `:`) takes two values – a value of type a and a `List a` – and creates a list from them. Here we are telling Haskell, *If our "length" function receives a value that is constructed with the help of ":" constructor, take it apart into constituents and do something with them.*

We have just defined our first *polymorphic* function – it works on lists of *any* type of values! We don't care what kind of value is stored in our list, we only care about the list's *structure* – you will see this as a recurring theme now and again, and this is another powerful aspect of typed functional programming. You define functions that work on types created from other "argument" types regardless of what these argument types are – you only care about the structure that the type function gives to a new type. Just like a "circle" is an abstraction of the sun, a vinyl record, and a coin, a type function is an abstraction of a lot of different potentially not very related entities. If you already started learning Haskell and got stuck on monad tutorials, this might be exactly the issue, being misled by false analogies. A circle is not the sun or a vinyl record, but it captures some common quality of both – they are round, even though at the same time the sun is hot and huge and a vinyl record can make you feel sad. The same with Monads – it's merely an abstraction of some useful quality of very, very different types. You need to learn to see this abstraction, without being lost in and confused by analogies, recognize it as such, and it will all make sense.

CHAPTER 2   TYPE CONSTRUCTION

Map a Function Over a List of Values

Do you recall a mapping example from the last chapter? Here's how we can define the extremely useful map function:

```
map :: (a -> b) -> [a] -> [b]
map f [] = []
map f (x:xs) = (f x):(map f xs)
```

The type signature tells us that map takes a function that converts values of type a to values of type b, whatever they are, and a list of values of type a ([a]) and gives us a list of values of type b ([b]) by applying our function to all values of the first list. It is likewise defined recursively with the boundary case being an empty list []. When map gets a value that is constructed with the help of : from x of type a and xs - remainder of the list - of type [a], we construct a new list by applying f to x and gluing it together with recursively calling map f over the remainder of the list.

Let us expand by hand what is going to happen if we call map (+2) [1,2]. (+2) is a function from **Int** to **Int**[1] - it takes any number and adds 2 to it. We want to apply it to all the elements of the list [1,2] - if you recall, internally it is represented as 1:2:[]. So

```
map (+2) (1:2:[])
-- pattern match chooses the case of (x:xs), with x = 1 and xs = 2:[]
((+2) 1) : (map (+2) (2:[]))
-- expanding recursive map call, we again pattern match the same case, expanding further:
((+2) 1) : ((+2) 2) : (map (+2) [])
```

---

[1] Strictly speaking, it is also a polymorphic function that works on any members of the Num typeclass, but we will get there.

```
-- expanding the last recursive map call, we match the "[]"
now, and get the final result:
(+2) 1 : (+2) 2 : []
```

The end result of such a call is what's called a *thunk*. In GHC, it will actually be stored this way until the actual values of the list will be required by some other function, e.g., showing it on the screen. Once you do that, GHC runtime will calculate all the sums and will produce the final list as [3,4]. This is part of Haskell being lazy and using the evaluation model "call by need." More on it later.

Dwell on the expansion above a bit, and it should become very clear and easy to use. So much so that you can work on some more exercises.

FUNCTION EXERCISES

Define some more functions:

1. "head" that returns the first element of the list.

2. "tail" that returns the remainder of the list.

3. "zipWith" that takes two lists of types "[a]" and "[b]" and a function f :: a -> b -> c that takes a value of type "a" and a value of type "b" and returns a value of type "c" and creates a new list of type "[c]" where each value is the result of applying "f" to corresponding elements of the first two lists. The type signature to help you a bit: zipWith :: [a] -> [b] -> (a -> b -> c) -> [c].

4. Write safe variants of all these functions using **Maybe**.

CHAPTER 2 TYPE CONSTRUCTION

One other piece of Haskell goodies that is worth noting in our card dealing code:

fullDeck = [Card x y | y <- [Clubs .. Hearts], x <- [Two .. Ace]]

It is called a *comprehension*, which might be familiar to you from math, and it basically tells Haskell to build a list of **Card x y** from all possible combinations of x and y from the right side of |. [Clubs .. Hearts] creates a list of all values of type **CardSuite** starting with Clubs and until Hearts, the same for [Two .. Ace] for values of type **CardValue**. You can obviously use the same syntax for **Int**: [1 .. 100] would create a list of all values from 1 to 100. You can in fact use this syntax with all types that belong to so-called Enum typeclass – that's what the deriving (Enum) statement is about, it makes sure we can convert values of our type to ints (details about typeclasses in the next chapters). Practice writing some comprehensions yourself, e.g., for pairs of int values, just pretend you want to put some dots on the screen and need to quickly create their coordinates in a list.

Info You can also create infinite lists – try this in ghci REPL (might not work online): "[1 ..]". It just keeps going forever, like the Energizer Bunny. Then, try this: "head [1 ..]" or "take 5 [1 ..]". We just applied a function to an infinite list and it worked fine, because Haskell is *lazy*. Laziness is *extremely* useful and beneficial in a lot of real-world scenarios, in some of which it makes Haskell outperform C, but they are more subtle and we won't be looking at them for quite some time yet, as well as what *lazy* and *strict* in fact mean.

Okay, so here's the full text of the card dealing program, which you should now be able to fully comprehend:

```
import          Data.List
import          System.IO

-- simple algebraic data types for card values and suites
data CardValue = Two | Three | Four | Five | Six | Seven | Eight |
Nine | Ten | Jack | Queen | King | Ace
  deriving (Eq, Enum)
data CardSuite = Clubs | Spades | Diamonds | Hearts deriving
(Eq, Enum)

-- our card type - merely combining CardValue and CardSuite
data Card = Card CardValue CardSuite deriving(Eq)

-- synonym for list of cards to store decks
type Deck = [Card]

instance Show CardValue where
  show c = ["2", "3", "4", "5", "6", "7", "8", "9", "10", "J",
"Q", "K", "A"] !! (fromEnum c)

-- please make sure to enable utf-8 in your terminal, otherwise
you may get errors when trying to execute
instance Show CardSuite where
  show Spades   = "♠"
  show Clubs    = "♣"
  show Diamonds = "♦"
  show Hearts   = "♥"

-- defining show function that is a little nicer then default
instance Show Card where
  show (Card a b) = show a ++ show b
```

CHAPTER 2   TYPE CONSTRUCTION

```
-- defining full deck of cards via comprehension; how cool is
that?! :)
fullDeck :: Deck
fullDeck = [ Card x y | y <- [Clubs .. Hearts], x <- [Two .. Ace] ]

smallDeck = [Card Ace Spades, Card Two Clubs, Card Jack Hearts]

main = print smallDeck >> putStrLn "Press Enter to deal the
full deck" >> getLine >> mapM_ print fullDeck
```

We looked at how the card deck types are constructed in the beginning of the chapter in detail. Keyword `data` is used to define new types and new type functions, as we already learned. Keyword `type` defines a so-called type synonym. In this case we write `type Deck = [Card]`, and then we can use `Deck` everywhere instead of `[Card]` - so it's mostly a convenience syntax for better readability, not so much useful here, but when down the line you will write `State` instead of `StateT m Identity` or some such, you'll come to appreciate it.

Syntax `instance <Typeclass> <Type name> where ...` is used to make a type part of a certain *typeclass*. We will look at typeclasses in detail in the next chapters, but a basic typeclass is similar to an interface in Java (they can be much more powerful than Java interfaces, but that's a future topic). A typeclass defines the type signatures and possibly default implementation of some functions, and then if you want to make a type part of a certain typeclass, you have to implement these functions for your type. Haskell also provides a powerful *deriving* mechanism, where a lot of typeclasses can be implemented automatically for your types. In the code above, we are deriving typeclasses `Eq` - which allows comparing values of our type - and `Enum`, which allows to convert values of our type to **Int,** and we hand-code typeclass `Show` for the types **CardValue, CardSuite,** and **Card.** We could have asked Haskell to derive `Show` as well, but then our cards wouldn't look as pretty. Try it out - delete the `instance Show...` lines and add `deriving (Show)` to all types and run the program again!

CHAPTER 2 TYPE CONSTRUCTION

A typeclass is another convenient and powerful mechanism to support abstraction and function polymorphism: without it, once you have defined a function `show :: CardValue -> String`, it would be strictly bound by its type signature. Then if you tried to define a function `show :: CardSuite -> String`, converting a *different* type to a string with a function of the same name, our **typechecker** would complain and not let you do that. Luckily, there's a typeclass `Show` that defines a polymorphic (working on different types) function `show :: a -> String`, which lets us work around this limitation. A lot of the power of Haskell in the form of Monads, Arrows, Functors, Monoids, etc. is implemented with the help of typeclasses.

Further down, the `fullDeck` function creates a full 52-card deck with the help of a list comprehension, and `smallDeck` is just three random cards thrown together in a list.

Then, our `main` function follows an already familiar logic of *break the problem down into smaller pieces and connect functions together making sure the types fit*. We are showing `smallDeck`, waiting for a user to press "Enter," and then mapping `print` over all elements of the `fullDeck` list to "deal" the full deck. `mapM_` is almost like `map` that we already defined – it maps a function over the list elements. The difference is that `print` is not really a mathematical function with the likes of which we dealt with so far – it does not map a value of one type to a value of another type, it puts something on the screen via calling an operating system procedure that in turn manipulates GPU memory via a complex cascade of operations. That's why `mapM_` is constructed differently from `map` – even though its high-level purpose is the same. We call such kind of manipulation "side effects." Side effects mess up our cute perfect mathematical world, but the real world *is* messy and all real-world programs are about side effects, so we *must* learn to deal with them!

As we shall very soon see, just like the sun, a vinyl record, and a coin are all round, **Maybe a**, **List a**, procedures with side effects, and a bunch of other very useful type functions have something in common as well – something that we will abstract away and that will make writing efficient and robust Haskell code for us an enjoyable and elegant endeavor.

Conclusion

In this chapter, we learned what algebraic data types are and how to construct them, how to build types from other types, how to define functions and data structures recursively, and how to start building polymorphic functions with the help of typeclasses. Now we can gradually move to more advanced Haskell topics, but before we do, let us look at a gentle Type Theory and Category Theory introduction to help structure what we've learned so far!

CHAPTER 3

Very Gentle Type Theory and Category Theory Intro

We will give some more strict Type Theory and Category Theory introduction in this chapter as it relates to Haskell. We will not focus on proofs too much (or at all) but will be providing (some) definitions and a hierarchy of concepts, which you can use as reference material as you delve deeper into the Haskell world. Strictly speaking, Haskell does not use the full intuitionistic Type Theory per se but is rather based on so-called System FC, or System F with coercions. The reasons for it are mainly that System FC, while being quite powerful (much more powerful than any imperative language), allows for fully deducible typechecking. This means you may omit writing type signatures altogether, and the compiler will deduce them for you. Try doing it in C# or Java.

However, we have found that knowing a more complete context of Type Theory helps reason about more advanced Haskell abstractions as well. Plus, in case you may want to venture to Idris, Coq, or invent your own language, the bigger picture will be even more helpful.

Also, one of the common misconceptions is that Haskell does not have so-called dependent types. Strictly speaking and from the type-theoretical point of view, it does – for instance, typeclasses and type families are nothing less or more than a so-called Σ-type, which is, in fact, *dependent*. So understanding some basic Type Theory will prove useful for any serious Haskell student.

As for Category Theory, well, Functors, Applicative Functors, Monads, and Arrows – the foundation upon which any real-world Haskell program is built – all come from Category Theory. And while you don't need to learn it at all to be a good Haskell programmer, knowing at least some basics will help make you *great* – appreciating the beauty of the concepts involved comes as an added bonus.

Types and Functions

We have already defined a **type** intuitively as "a bunch of stuff with similar characteristics thrown in a pot together." In intuitionistic Type Theory types are usually referred to with capital letters – such as A, B, etc. In Haskell, concrete types (such as **Int**, **Char**, etc.), as well as *type functions* (we will define them strictly below), that take a type and construct a different type out of it, such as **Maybe a**, have to start with a capital letter as well.

In math, we construct all types from nothing, starting with natural numbers N, which are usually called Nat in programming languages. One of the ways to construct natural numbers in Type Theory is as follows:

$$0 : N$$

$$succ : N \to N$$

We postulate the existence of zero as a value of type N and define a function from N to N. This way, we can say that $1 \equiv succ\ 0, 2 \equiv succ\ 1$, etc. In a similar vein, we can construct integers, reals, rings, fields, and basically all of math.

CHAPTER 3   VERY GENTLE TYPE THEORY AND CATEGORY THEORY INTRO

In Haskell (*or any other real-life language for that matter*), we have a bunch of primitive types already provided to us as a basis from the operating system or a specific compilation target, upon which we build our Type Universe using a subset of tools and rules detailed below.

> **EXERCISE**
>
> Define natural numbers in Haskell.

Functions between types are written as $f: A \to B$, a notation that denotes a function with *domain* equal to type A and *codomain* equal to type B. In Haskell, the notation is very similar, with the only difference being that Haskell uses a double colon for the type signature. The function f above would be written as f :: A -> B, or recalling an example from the previous chapters (see Figure 3-1), function whichNumberIsChar :: Char -> Int.

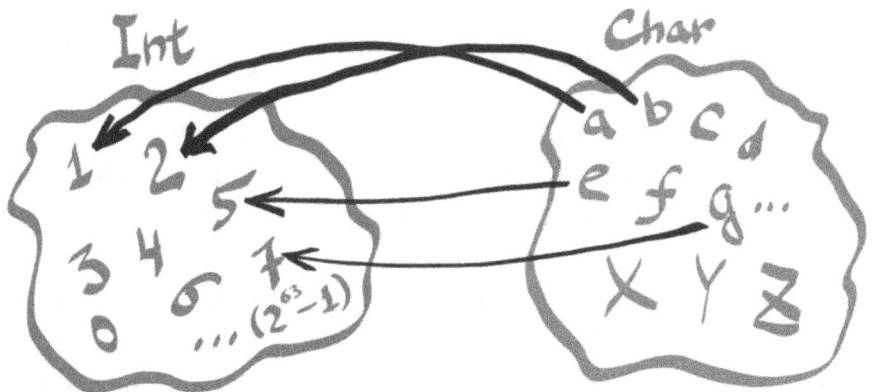

Figure 3-1. *"whichNumberIsChar"*

This notation also means that if we have a value a of type A, denoted as a:A, we can apply a function $f: A \to B$ to it and will get a value f(a):B - f(a) of type B. As an example, having a value "Z" of type **Char**, we can apply a function `whichNumberIsChar` to it and receive a number – a value of type **Int**. In both Type Theory and Haskell, function application does not require parentheses and is written simply as `f a`.

Important to note that functions are also first-class citizens of Type Theory, which means we can pass them around as values, apply other functions to them, and classify them in types. So $A \to B$ is a type of *all* functions from A to B, just as *Char* $\to$ *Int* is a type of *all* functions from Char to Int.

Types in Type Theory are categorized in the hierarchy of *universes* denoted with the letter U. So, in a sense, the "type" of a type is a universe – we can write `A:U` or `Char:U` to explicitly denote this fact. In math, universes are organized in a hierarchy $U_0 : U_1 : ...$, but it's not really relevant for Haskell. By definition all types that belong to a universe U_n also belong to a universe U_{n+1}, so we will simply write `a:U` without referencing a number for the universe explicitly, when we want to talk about a type.

Maybe and Advanced Generalized Functions

So far, we have described concrete types where certain objects – values – live as well as functions between them, converting a Char to an Int or an Int to an Int with something like `f x = x*x` and so on. But what other "arrows" can we introduce in our universe?

One example that we have already seen is what we've called a "type function" – a function that takes a type and creates another type from it. We have defined two new types this way – **String** from **Char** and **ListInt** from **Int**. Since the type of any concrete type is a universe U, the type of *all* such "type functions" is `U -> U`. In Haskell, when we talked about a type variable a, we used to write `a:*` where "*" denoted a concrete type, but lately the accepted way became to write `a:Type` instead. Haskell also has a so-called kind system, which helps distinguish between different type functions basically – e.g., Type is one kind (concrete type), `Type ->`

Type is another kind (a type function that creates another concrete type from some type), etc. This is actually a bit confusing and becomes much clearer if we think about kinds in frames of full Type Theory – as here it simply is a function, just the one that operates on types as values. A ubiquitous example is the two "type functions" we defined in the previous chapter – `Maybe a` and `List a`. Using GADT syntax, as it shows the relevant concepts better:

```
data Maybe a::Type where
        Just :: a -> Maybe a
        Nothing :: Maybe a
```

In effect, `Maybe : U -> U` is a function that takes *any* concrete type `a:U` and creates a *new* concrete type `Maybe a` out of it. We can even write in Haskell `type MaybeInt = Maybe Int` where we give a new name – MaybeInt – to the type that we get by applying a type function Maybe to a concrete type Int. In practice, no one really does it with Maybe, as it's easier to just write `Maybe Int`, but you may want to introduce such named new types with more complicated type functions such as monad transformer stacks, which we will look at much later.

A very important question is what is `Just` and `Nothing` in our case? Let's stare at the type signatures attentively. We can rewrite `Just` as `Just(x:a) : Maybe a`, which means that `Just` takes a *concrete value* of type a and returns a *concrete value* of type `Maybe a` as a result. It is also called a constructor function. Same with `Nothing` – it takes no arguments and returns a value of type `Maybe a`, or we can even say that `Nothing` *is* a value of type `Maybe a`.

You have to internalize this distinction: `Maybe` is a "type function" – it takes a *concrete type* as an argument and returns a new *concrete type* as a result. Under the hood, it combines two data constructors – `Just` and `Nothing`, which can be treated as "normal" functions of type `a -> Maybe a`.

CHAPTER 3   VERY GENTLE TYPE THEORY AND CATEGORY THEORY INTRO

So what happens if we do write `type MaybeInt = Maybe Int`? You simply need to substitute type variable a for the concrete type `Int` in our `Maybe a` definition:

```
data Maybe Int where
        Just :: Int -> Maybe Int
        Nothing :: Maybe Int
```

So our type function `Maybe` got a specific type (`Int`) and produced a new concrete type with two data constructors as a result. This is exactly what happens under the hood when you write something like `x = Just 4` – Haskell deduces that 4 is of some concrete numeric type (can be Int, Short, etc.), creates a concrete type Maybe Int from Maybe a based on this information, and then applies a data constructor `Just :: Int -> Maybe Int` to our `4::Int` value.

Here we can also note an important difference between U in Type Theory and "*" or Type in Haskell. The latter only means a *concrete* type, but cannot mean a type function such as Maybe itself or List – i.e., you cannot write `Maybe List` in Haskell, but you can write `Maybe (List Int)`, since `List Int` is a concrete type and just `List` is a type function. In Type Theory, however, both Maybe and List have a type U since they live in our universe of types.

So far, we have looked at "regular" functions – which take a value of some type as an argument and return a value of some other type – and "type functions," which take a concrete type as a value and return some other type (such as `Maybe a` or `List a`). Both these kinds of functions are present in Haskell. But we can also conceive of two other kinds: functions that take some type as an argument and produce a value of some type and functions that take some value of some type as an argument and produce a new type. Here they are summarized in a list and a scheme to help you structure these notions (Figure 3-2):

56

- "Normal functions": a -> b in Haskell. Take a value of type a as an argument and return a value of type b as a result.

- "Type functions": Type -> Type in Haskell. Take some concrete type as an argument and return another concrete type as a result. The main mechanism to construct new algebraic data types (ADTs) in Haskell.

- Functions from types to values: Type -> a in Haskell syntax. Take some concrete type as an argument and return a value of some type a as a result. At first glance, they don't really exist in Haskell. However, when we introduce typeclasses formally, we'll see some similarities.

- Functions from values to types: a -> Type in Haskell syntax. This is what people normally understand by "dependent types" and what is again strictly speaking not *fully* available in Haskell, but there is very active research and some GHC extensions allow some types (such as String or Nat) to be used in limited dependent functions. Also, see comments in the "Dependent Function Types (Pi-Types)" section.

CHAPTER 3   VERY GENTLE TYPE THEORY AND CATEGORY THEORY INTRO

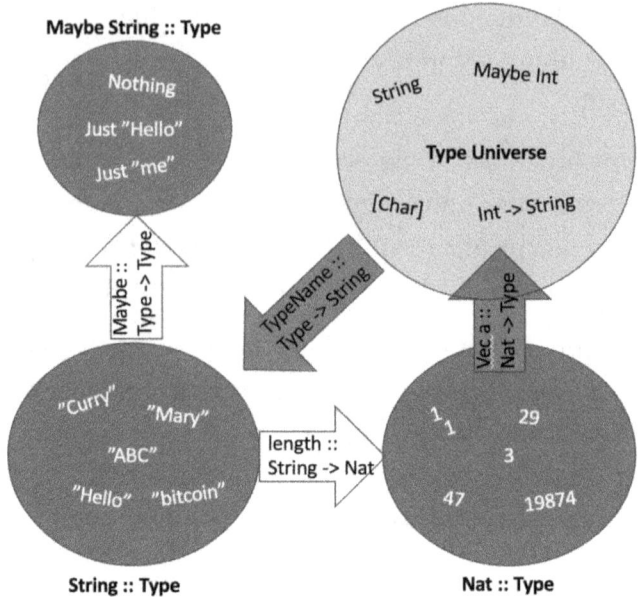

Figure 3-2. *Four types of functions: from values to values, from types to types (both present in Haskell), from values to types (Vec a), from types to values (TypeName) (these are not really present in Haskell, with some experimental exceptions as of the time of this writing). Also, as discussed below, it is possible to interpret Haskell typeclasses as functions from types to (function) values*

How about functions of type U -> A or Type -> a, i.e., functions that take a concrete type as an argument and produce a value of a concrete type as a result? For example, what if we want to have a function EN : U -> Nat that enumerates all the types that live in our universe and returns a specific type's number? We could say that EN Char = 0, EN Int = 1, etc. This is a perfectly valid construct in Type Theory, but not in Haskell. However, when we look at sigma types further below, you will see that it is not entirely true.

What if we go the other direction and want a function of type A -> U or, equivalently, a -> Type? This is what is normally called "dependent types" in CS, or *indexed type families* in Type Theory, and is implemented

58

in languages such as Idris, but not strictly speaking in Haskell (with exceptions of some polymorphic functions as discussed below), even though as mentioned above there is active research in this direction. Everyone's favorite example is the type of vectors of a given length n. Ensuring safety of a bunch of operations that require vectors of the same size would work much better if we could do it at type level rather than by introducing runtime checks. This topic brings us to the following.

Dependent Function Types (Pi-Types)

A dependent function type is a kind of functions, whose codomain (type of their result) depends on the argument. It is formally written as

$$\prod_{x:A} B(x)$$

and means that a function of this type takes an argument x of type A and returns a result of type B(x) - so a type may vary depending on what x is. In case the type B(x) does not depend on x and is constant, it becomes our "regular" function type A -> B.

As an example, let's say we have a type Vec n:Nat a:U of vectors of a given length n that store elements of type a. For example, we can use Vec 3 Double to store coordinates of physical bodies or Vec 1536 Double to store vector embeddings of the text for AI large language models. Then, we should be able to define a concatenation function that takes two vectors of types Vec n:Nat a:U and Vec m:Nat a:U and produces a vector of *new* type Vec (n+m):Nat a:U that depends on the exact type of its arguments. The full type signature of such a function would be written as

$$\prod_{x:(Vec\,n:Nat\,a:U)} \prod_{y:(Vec\,m:Nat\,a:U)} Vec(n+m):Nat\,a:U$$

CHAPTER 3   VERY GENTLE TYPE THEORY AND CATEGORY THEORY INTRO

This is a mouthful. In this specific case since the result type depends on the *type* of the arguments, not on the arguments themselves, we can actually use our "regular" function signature type and write Vec n:Nat a:U -> Vec m:Nat a:U -> Vec (n+m):Nat a:U – which is in fact quite similar to how it is done in Idris.

As mentioned, pi-types in the sense of the example above have very limited and quite experimental support in Haskell as of the time of this writing, so we will not spend much more time discussing this. However, there is one important exception that's worth noting. Turns out, virtually all polymorphic Haskell functions are in fact pi-types! Consider the simplest Haskell function, id:

```
id :: a -> a
id x = x
```

This is a short type signature. The full type signature for this function in Haskell is forall a. => a -> a. It works for *any* type a and returns whatever it was given, regardless of what concrete type a is: it can be Int, String, some function type, etc. But this also means that the codomain of this function depends on the domain of this function, which is exactly the definition of pi-types! The type signature for the id function in full Type Theory is

$$\prod_{A:U} A \to A$$

which in essence means that the type of both the domain and codomain of this function depends on one type variable A:U. Haskell takes a shortcut and makes the need for us to instantiate type variable a implicit using the forall a quantifier. Under the hood, when you write something like id 4 in Haskell, it gets unwinded into the full form – id 4::Int – then our type variable a gets instantiated with a concrete type **Int** and only then typechecking is done, id :: Int -> Int 4::Int = 4 :: Int, and all is good.

As another example, let's take a function `length :: forall a. => [a] -> Int` that measures length of any list and that we defined in the last chapter, and again, it works independently of what concrete type `a` is instantiated to, hence the `forall` quantifier. In Type Theory, the full formal type signature for this function is

$$\prod_{A:U}[A] \to \text{Int}$$

This tells us that when we apply this function to a list of specific value types, we first instantiate the type variable `a` to this specific type and only then do everything else.

So, despite Haskell not having full-featured pi-types (or dependent functions) in a sense that the type of the function's codomain cannot depend on the value of a specific type (such as 4::Nat or "hello"::String), it *can* and *does* depend on type values (such as Int::Type or String::Type) in polymorphic Haskell functions.

Sum, Product, and Dependent Pair (Sigma) Types

We have informally introduced sum and product types in the previous chapter when discussing ADTs. Sum types are also called coproduct types and are defined as: given two types `A,B:U` we can introduce the type $A + B : U$. We won't go into formalities here, and specifics on how to use sum types in Haskell are discussed above. For product types, given two types `A,B:U` we can introduce the type $A \times B : U$ that is called their *cartesian product*. Elements of this type are pairs (`a:A, b:B`). In the previous chapter, we discussed in detail how product types can be introduced via the `data` keyword either as unnamed tuples or as records with named fields. When we have introduced pairs, moving to any number of elements in the tuple is trivial via successive application of the "pairing" operation.

CHAPTER 3   VERY GENTLE TYPE THEORY AND CATEGORY THEORY INTRO

Things get more interesting when we generalize our product types to so-called dependent pair, or sigma, types. If we have a type `A:U` and a *family* `B:A->U` (which is again a function from values of our type A into universe U, i.e., its results are concrete types), then a dependent pair type is

$$\sum_{x:A} B(x)$$

Constructing elements of this type is done – just as for an "ordinary" product type above – via pairing: `(a:A, b:B(a))` are the elements of the sigma type above. This is also where generalization becomes apparent: if B does not depend on a, then we simply have `(a:A, b:B)` – a boring normal product type.

Sigma types are ubiquitous in Haskell, even though they are never called that – but in fact all the typeclasses and type families are sigma types! Haskell somewhat confuses matters here, by treating typeclasses, multiparameter typeclasses, type families, data families, and type synonym families as separate entities, while in fact *all of them* are nothing more (or less!) than various sigma types. As we introduce these Haskell concepts, we will come back to this point repeatedly to help you clarify and structure things.

The next chapter introduces many examples of typeclasses. For now let's take a look at a simple example from abstract algebra and define a structure called "Magma" – which is basically a type with a binary operation defined on it:

$$\text{Magma} := \sum_{A:U} \circ : A \to A \to A$$

Magma is a foundation for more complex algebraic structures, such as Monoids, Groups, Rings, etc., and is interesting because given a type `A:U` it also defines a certain structure for this type, namely, the ability to combine two elements of this type to receive a third one. All number types, strings, all lists, and many, many others are Magmas (they of course have much deeper structure as well – Magma is just probably the simplest one we can come up with). More complex sigma types allow us to define more complex structures or relationships between different types and values, and as soon as we start thinking about relationships between objects, we inevitably come to the need to look at some very basic Category Theory.

Very Gentle Category Theory Introduction

We have talked quite a lot about function composition in the previous chapters as the essence of functional programming. Well, composition per se is the essence of Category Theory – so there must be some relation, right? Turns out, there is very close relation, and many of the key Haskell abstractions – such as Functors, Applicative Functors, Monads, and Arrows – do indeed come from Category Theory. While you can use them proficiently in practice without knowing about this connection, learning about it does help have a much deeper understanding of the concepts involved and will help you eventually see these (and other) abstractions in the areas you couldn't identify them before – which makes you a more efficient and better architect, programmer, and math practitioner.

Category Theory deals with objects and relationships between them, shifting focus to the relationships from the objects themselves. This approach proved to be very fruitful in mathematics and found many applications in very practical areas, such as computer science, business process modeling, databases, etc. Let's see how it relates to Haskell.

Category C (see a very simple example in Figure 3-3) is anything that satisfies the following characteristics:

- A bunch of objects – formally, a (mathematical) class, but in our case a set will suffice.

- Arrows, also called morphisms, or maps, between these objects. We will mostly use morphisms when talking about categories in general, since Arrows are also a very important Haskell typeclass and map is a ubiquitous function. Every morphism has a source object a and a target object b, and we will write `f:a->b` for such a morphism.

- There must be an identity morphism for every object, $id_a : a \to a$.
- Morphisms have to compose: if we have morphisms `f:a->b` and `g:b->c`, then their composition is written as $g \cdot f$ and it turns a into c: `a -> c`.
- Morphism composition must be associative: $h \cdot (g \cdot f) = (h \cdot g) \cdot f$.

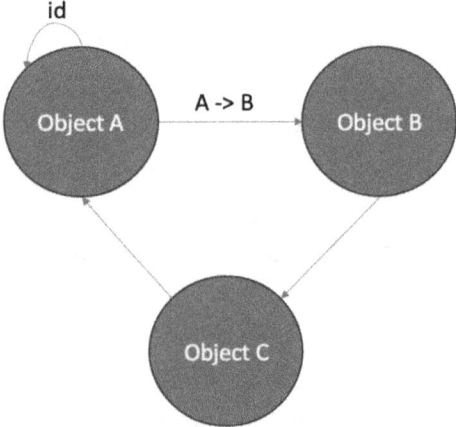

Figure 3-3. *Category with some morphisms shown*

That's all there is to it! If all of it seems very familiar, it indeed is – if we take types as objects and functions as morphisms, we get a category! Functions compose, they are associative, and there is identity function. In mathematics, the most studied category is category **Set** – category of sets and functions between them. On one hand, Haskell should fully correspond to it, but there is one caveat: Haskell has a type called "bottom," noted as symbol ⊥ and whose meaning is "a computation that never completes successfully." Its addition, along with some advanced questions about the definition of morphism equality in Haskell, makes category **Hask** different from **Set**, and some even question whether it is a category depending on the definitions used. However, for all practical purposes it doesn't really matter, and we can freely reuse a lot of categorical knowledge when thinking about Haskell.

> **CATEGORY EXAMPLES**
>
> Come up with some examples of what may constitute a category. Any Graph is definitely one. What else?

So why is at least some categorical thinking helpful when building practical Haskell programs? Recall how we approach solving the problems in the typed functional world:

- Analyze your problem domain and design the types that you will need.

- Decompose the problem into progressively smaller pieces, solvable by functions that compose (the input type fits the output type of the previous one!).

Everything starts with designing the types – objects of the category specific to your program (*since Hask is a category, (almost) any program written in Haskell will be its subcategory; no, we will not prove it here*). Then you design your morphisms – functions between your types. In essence, when you write a Haskell program, all you do is create a new category based on **Hask**! This means that the more you are familiar with the categorical way of thinking and with Haskell abstractions that come from Category Theory, the more efficient and beautiful your programs will become.

The second, subtler, point is that the *essence of any category is composition*. This is exactly what we want from our program – if our functions compose, it will compile. If it compiles, it will run correctly in most of the cases (barring the variables we don't control – such as erroneous user input and others).

The main instruments through which Haskell introduces abstractions and categorical notions are typeclasses (and their generalizations – various type families). Arguably, the main Haskell typeclass hierarchy is **Functor–Applicative–Monad**, and we will study it in detail in further chapters. Then there are **Arrows**, a relatively recent addition to Haskell, which in a sense generalizes monads as computation abstraction even further and is based directly on **Category** typeclass (yes, there is one). We will study it as well. For now, let's introduce typeclasses more or less formally, and let's look at how they are connected to both sigma types and Category Theory.

Typeclasses

The simplest form of a typeclass is a typeclass dependent on one type parameter. It can be introduced in the following way:

```
class <Name> a where
    func1 :: a -> <SomeType>
    func2 :: a -> a
    func3 :: <SomeOtherType> -> a
```

For example, Magma that we defined type-theoretically above can be written as follows:

```
class Magma a where
    (*) :: a -> a -> a
```

```
-- and, if we want to make let's say type Int an instance of
Magma, we write:
```

```
instance Magma Int where
    (*) = (+)
```

Above we have defined a Magma typeclass and made a type Int an instance of this typeclass, cheating somewhat as we used an already defined operator (+) as the body of the Magma operator instead of building it "from scratch." In reality, it's a bad idea to redefine (*) as (+) since (*) is multiplication for all numerical types, but one of Haskell's powers is that you can use pretty much any symbol sequence as an operator, including all of Unicode, so feel free to experiment.

By looking at the above code, we can see that a typeclass defines a type variable a that can be instantiated to a concrete type and then *names* and *type signatures* of the functions that this typeclass contains, where type signatures should contain at least one occurrence of the a type variable. That's where similarity with sigma types should become apparent – basically, a one-parameter typeclass defines the following:

$$\text{typeclass} := \sum_{A:U} \langle \textit{list of functions with A in the type signature} \rangle$$

So a basic typeclass defines a type signature for a dependent sigma type, where the only allowed dependence is on one type variable a. That is also why we can interpret typeclasses as functions from types to *function values* – given a specific type it provides specific functions (which are also values!) that work with this type. When you want to add the structure defined by your typeclass to a certain type, you write an *instance* of this typeclass for this type – similar to what we did to Int and Magma above.

Another very basic example of a one-parameter typeclass is Show:

```
class Show a where
    show :: a -> String
```

It defines one function from type a to String and is used primarily in debugging or as a quick-and-dirty serialization mechanism for any Haskell type. For many Haskell typeclasses, many functions (and for some all functions, so all the typeclass instance) can be *derived* automatically – and you will appreciate this fact many, many times when writing Haskell programs. Show is one such typeclass – you do not need to write instances of Show for your custom types, but can simply use the *deriving* keyword:

```
data Person = Person {
      name::String,
      age::Int
  } deriving (Show)
-- and then you can write with 'show' being defined
automatically by the compiler:

s :: String
s = show $ Person "John" 42
```

From a categorical point of view, a one-parameter typeclass either adds some internal structure to a type A by defining an *endomorphism* (such as a Magma binary operation – think addition, multiplication, function composition, etc.) or creates some structure for your program category, such as a Show typeclass that creates a morphism between any type A and a String type (remember that types are objects in Hask).

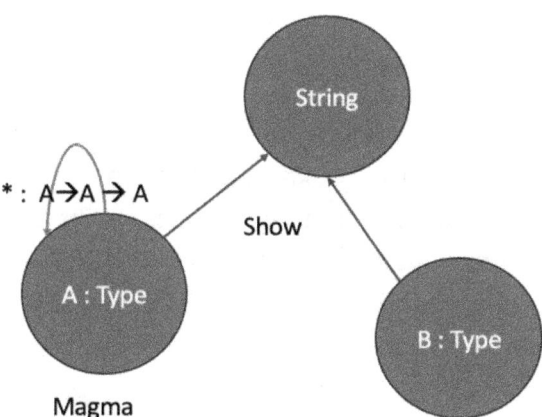

Figure 3-4. *Magma typeclass adds internal structure to a type (by providing endomorphism – binary operation A→A→A). Show typeclass adds structure to the whole category, by creating relations between types (A and B in the Figure 3-4) with String type*

In Haskell (at least the most popular and the de facto standard GHC), you can also use multiparameter typeclasses that extend the dependency of functions to several type variables:

$$\text{multiparamtypeclass} := \sum_{A,B,C,\ldots:U} \langle \text{list of functions with } A,B,C\ldots \text{in the type signature} \rangle$$

These typeclasses define certain morphisms *between several types* and as such enrich your program category with additional structure patterns between objects. In practice, this means that once you see the pattern of relationships between several different types that can be applied to other types, you have a candidate for a multiparameter typeclass. They come handy in different data structures, but not only, as we will see.

One last thing to note: In Haskell typeclasses' type variables can be not just concrete types, but also *type functions* (such as Maybe a or List a that we are already familiar with). This can be made explicit by adding type signatures to the type variables in typeclass definitions:

CHAPTER 3   VERY GENTLE TYPE THEORY AND CATEGORY THEORY INTRO

```
class Functor (f :: Type -> Type) where
class Show (a :: Type) where
```

With typeclasses and their generalizations such as type families, which we will describe further in the book, Haskell becomes extremely powerful – even without full support of pi- and sigma types we discussed above. To recap, when designing your program, think about it in categorical terms: types are objects, morphisms are functions, and typeclasses capture internal type structure as well as additional structure between objects in your category. This way your programs will become concise, beautiful, efficient, and easy to maintain.

Conclusion

In this chapter, we looked at some of the theoretical foundations behind Haskell and all other cutting-edge computer science research – Type Theory and Category Theory. We will apply this knowledge in the next chapters to have more structured and solid approach to solving problems in the functional way.

CHAPTER 4

Basic Typeclasses or "Show Me a Monoid"

- "What's your name?", you ask.

- She looks at you in amazement, but with warm concern. "You really don't remember anything?"

- "I ... It starts to come back to me slowly – like I could deal those cards now – but it's all very vague still ..."

- "Well, not to worry. I will guide you! And by the way, my name is Lambda, and we are friends!", she says after a slight hesitation. *Lambda and Curry*, you think to yourself. Sounds strangely fitting. "Let me help you remember more of your magic, Curry! You'll need it for the contest ... Let's start with discussing how the same spells can do different things depending on the types you apply them to!"

The idea of some "contest" gives you another reason to worry, but you decide to learn as much as you can for now.

Let us recap our high-level **Hask** picture so far. We have *values* grouped together in *types*, such as Char or Int. We have *functions*: mapping of one, or several, values of one type to one and only one value of a type, such as length :: String -> Int or fact :: Int -> Int. Functions are

CHAPTER 4   BASIC TYPECLASSES OR "SHOW ME A MONOID"

values, too – you can pass them to a function and define types consisting of functions as values. We have *type functions* or *parametric types* – they take a *whole type* as a value and construct a different type out of it. Such as a list type **[a]** or **Maybe a**, where variable a can be of any concrete type. We have *polymorphic* functions – those that are defined on parametric types and work regardless of what exact type a is, such as `length :: [a] -> Int` or `map :: (a -> b) -> [a] -> [b]` that applies any function `f::a->b` to all values of the list **[a]**.

The real power (and complexity) of Haskell comes from various abstractions, like polymorphic functions we looked at in the previous chapters. The next such mechanism is **typeclasses**.

What is a typeclass? If you read Chapter 3, you already know that typeclasses, multiparameter typeclasses, as well as type and data families are *all* simply cases of so called Σ-types, or dependent product types. In essence, they take a type (or several in case of multiparameter typeclasses) and add some structure to the type (or between types for multiparameter typeclasses). In the next few chapters, we will look at different kinds of important typeclasses using ubiquitous Haskell examples and try to show their increasing power of abstraction.

Show Typeclass

The simplest case of a typeclass is a single-parameter typeclass that works on concrete types (as opposed to type functions such as `Maybe a`). Such typeclasses are similar to Java "generics" or C++ "templates," but even the next kind of typeclass, described in the "Lift Me Up!" section, which works on *type functions* (with type signature such as `Type -> Type`) rather than concrete types, does not exist in widespread imperative languages, making them much less concise and powerful than they could have been.

An easy example of typeclass is **Show**:

```
class Show a where
    show :: a -> String
```

It defines a function show that takes a value of some type a :: Type and converts it to **String**. If we take our type **Person** defined previously, we can make it an *instance* of this typeclass as follows:

```
instance Show Person where
    show (Person name age) = name ++ " is " ++ show age ++
    " years old"
```

Notice how there are no implicit type conversions in Haskell, so we cannot simply write ... ++ age ++, since age is of type **Int**, not **String** that operator ++ expects. Otherwise, it's pretty straightforward – you define function bodies for your types that you want to make a part of a certain typeclass, and these functions start working on your types.

Typeclass **Show** defines a useful (and straightforward) abstraction "how should a value of any type be represented as a string?" In our type-theoretic notation from Chapter 3, **Show** would be represented as

$$\sum_{a:U} show : a \rightarrow String$$

which basically says "it's some type a with the function show :: a -> String defined."

Algebra Is Cool

Now let's get more abstract and consider some type A:U – a bunch of objects with some similar properties grouped together. It can be **Nat** (natural numbers) or **String** or **Char** or whatever else you may think of.

CHAPTER 4   BASIC TYPECLASSES OR "SHOW ME A MONOID"

Taken in itself, it's nothing more than that – a bunch of objects thrown together. We can't really do anything with them, there is no *structure*. What if we add some to make things more interesting?

Let us define some binary operation ∘ for our type, i.e., a function that takes two members of our type and turns it into a third: ∘ : A → A → A. It can be addition for **Nat** or integers, string concatenation for **String**, or operation of "having sex" for people, which sometimes tends to produce other people. A type with such a binary operation is called a **Magma** in algebra. Such an operation adds a certain *structure* to our type and makes it more interesting to work with – as now we can combine any two elements of our type and get a third one. If the abstraction of *type* can describe pretty much *everything* we can think of, the abstraction of *Magma* is much more specific – not every type allows introduction of such an operation – but still general enough to describe both natural numbers and humanity. If we continue adding *structure* via additional operations, functions, and laws on them, we gradually get more and more specific.

The art of programming is finding the right level of abstraction for our model so that it is general enough to use concise and powerful abstractions, yet specific enough to be able to correctly describe all the objects in our model and relationships between them. You can code anything in JavaScript or any other language by being very specific – but it tends to be verbose, difficult to understand and maintain, and error-prone. If you use good abstractions in a strongly and statically typed language such as Haskell, your code will be concise, efficient, fast, and easy to maintain.

In Type Theory, Magma is defined as

$$Magma := \sum_{A:U} A \to A \to A$$

which says exactly what we described above – it's a type A with a binary operation defined on it.

What if we want even more structure and require our ∘ operation to be associative, i.e., make sure that $x \cdot (y \cdot z) = (x \cdot y) \cdot z$ for all x, y, z? We turn our *Magma* into a *Semigroup*. It is easy to see that natural numbers with addition is a *Semigroup*, as well as **String** with string concatenation. However, people with operation "having sex" is definitely not a *Semigroup*, and the equation itself is against the law in most of the countries. I think we've taken this analogy far enough.

Let us also add something that is called an *identity element* to our *Semigroup* type and we get a *Monoid* – type A with an associative binary operation and an identity element e. For an identity element to be called that, we require that $x \cdot e = x$ as well as $e \cdot x = x$. For natural numbers with addition, an identity element is 0 and for strings with concatenation an empty string "". What is the identity element for natural numbers with multiplication?

Monoids are a cool abstraction because it has an interesting structure, but at the same time they describe a lot of very different types in Haskell – not just all the numerical types and strings, but also any lists and some special kinds of *computations*.

Think about one example of the generic usefulness of the Monoid typeclass with the types that we have already described.

Type-theoretically, Monoid can be defined as

$$Monoid := \sum_{A:U} \sum_{e:A} \sum_{\cdot: A \to A \to A} \prod_{(x,y,z):A} x \cdot (y \cdot z) = (x \cdot y) \cdot z, e \cdot x = x, x \cdot e = x$$

which summarizes everything we discussed above and gets really tedious to write down. But how do we define and use these cool Σ-types in Haskell?

MONOID EXERCISE

Try to write a typeclass for the *Monoid* definition before reading further. An answer to the previous exercise is, for instance, an ability to summarize a list of any type a::Type in Haskell that is a *Monoid*. Since we have an operation that combines two values of our type into one, we can apply it to all the elements of the list in order and get a final result. For a list of numbers, this will obviously be a sum (or product) of the list, but for other types with *Monoid* structure, it may mean something different, but always a natural way to *fold* a list of values. Now you can define such a polymorphic function – do it!

Typeclass Hierarchy in Haskell

In Haskell, we could define *Monoid* as follows:

```
class Monoid a where
        e   :: a
        (*) :: a -> a -> a
```

In reality, it is defined a bit differently, as typeclasses can extend each other – a very useful feature, since we can follow along our gradual structure buildup in the same logic as described in the previous section. That is, we could define a *Magma*, then extend it and get a *Semigroup*, and then add an identity element and get a *Monoid*. As of this writing, in the base Haskell libraries, *Monoid* extends *Semigroup* roughly as follows:

```
-- semigroup is simply a type with binary op
class Semigroup a where
        (<>) :: a -> a -> a

-- note the extension syntax when defining a Monoid:
class Semigroup a => Monoid a where
```

```
-- identity element
      mempty :: a

-- mappend has a default implementation and equals operator
(<>) from Semigroup
-- this is a legacy method and will be removed in the future,
as (<>) is available
      mappend :: a -> a -> a
      mappend = (<>)

-- mconcat is the list folding function we discussed above!
      mconcat :: [a] -> a
      mconcat = foldr mappend mempty
```

Here we notice several interesting things. First of all, there is unfortunately no way to put the mathematical *Semigroup* or *Monoid* laws in typeclass definition – Haskell has no way of enforcing or checking them. So typeclasses can only define *function type signatures* and *default implementations* for these functions. However, it is *extremely important* to make sure your types follow relevant typeclass laws that are always mentioned in the typeclass documentation if you make your type an instance of a certain typeclass. The reason for this is that the Haskell compiler (at least GHC) often uses these laws when running various optimizations – so if your type does not respect them, your code may start doing things that you do not expect.

Second, *Monoid* typeclass in Haskell has the mconcat function added with the default definition provided. This function folds a list of values of a monoidal type using its <> binary operation. The beauty of default definitions is that when you make your type an instance of *Semigroup* and then *Monoid,* you only need to define the binary op <> and the identity element mempty. Once you've done that, you can safely apply mconcat to lists of your type, and it will work "automagically." As we will see down the road, a lot of more complex typeclasses provide many more methods

CHAPTER 4 BASIC TYPECLASSES OR "SHOW ME A MONOID"

with default definition, which makes a programmer's life much easier – once you learn to use the typeclasses well. You can also provide your own definition for a default method if you think you can make it more efficient.

CARDS MONOID

Try to define a Monoid instance for our **Cards** type from Chapter 2. What would be the identity element? Does the associativity hold? What other Monoids can you think of in your life?

The other very powerful feature of typeclasses is that you can define instances for parametric types (or *type functions*) as well as concrete ones. For example, you can make sure that a list of any type [a] is a *Semigroup* by defining the binary operation to be equal to list concatenation:

```
instance Semigroup [a] where
        (<>) = (++)
```

Once you have this definition (and it is a part of the base library), you can use operator <> for any lists – e.g., [1,2] <> [4,5] should produce [1,2,4,5] as a result. If you are designing a library or a complex enough project, it really pays to think through the typeclass structure and define as many "automagical" instances as makes sense to help you avoid boilerplate and make code more readable and maintainable.

Now that we have seen a glimpse of what basic typeclasses can do for us, let's move on to typeclasses defined for the *type functions* or *parametric types* such as **Maybe a** or **[a]**.

CHAPTER 4 BASIC TYPECLASSES OR "SHOW ME A MONOID"

Lift Me Up!

Recall how when we were discussing safe division operation when introducing **Maybe** in one of the first chapters, we changed *this* for a function square x = x * x:

```
divideSquare :: Double -> Double -> Maybe Double
divideSquare x y = let r = divide x y in
                   case r of Nothing -> Nothing
                             Just z  -> Just (square z)
```

into *this*:

```
divideSquare :: Double -> Double -> Maybe Double
divideSquare x y = square <$> divide x y
```

This transformation hints that operator <$> somehow applies the function square::Double->Double to any value of type **Maybe Double** (since that's what divide::Double->Double->Maybe Double returns) using some sort of pattern matching or case statements.

> **MAYBE MAP EXERCISE**
>
> Before reading further, try to define a function maybeMap that would take *any* function f :: a -> b and would allow you to apply it to a value of type Maybe a and return Maybe b, in effect turning function f to function g :: Maybe a -> Maybe b.

What happens here is that we take a function between values of two concrete types (in case of square Doubles, but it can be anything) a and b and magically receive the ability to apply it to values of type Maybe a and return Maybe b. We can say that we took a function a -> b and *lifted* it into Maybe *type function*, turning it into a function Maybe a -> Maybe b. See Figure 4-1 for visualization of the concept.

79

CHAPTER 4   BASIC TYPECLASSES OR "SHOW ME A MONOID"

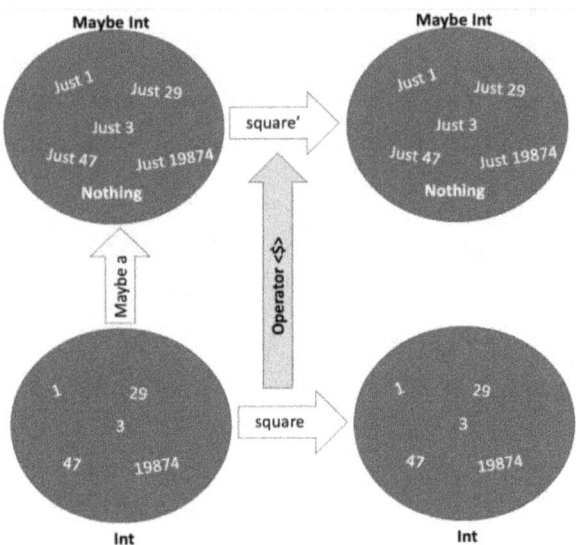

Figure 4-1. *Lifting Ints to Maybe*

By looking at this attentively and thinking hard, we can easily design operator <$> for Maybe like this (note also operator definition syntax):

```
(<$>) :: (a -> b) -> Maybe a -> Maybe b
(<$>) f Nothing = Nothing
(<$>) f (Just x) = Just (f x)
```

It takes a function between *some* concrete types a and b and turns it into a function between corresponding Maybes. Now, Maybe a is a type function that you should read as Maybe (a:Type) = ... and that takes a concrete type (a) and makes another concrete type out of it. What other type functions do we know by now? Of course it's a **List**, or **[a]**! Can you define operator <$> that would take a function a -> b and turn it into a function between corresponding lists – [a] -> [b]? I know you can. Do it. Do it now.

Good. As they say, one is a coincidence, two is a trend, so why don't we take it further? What if we had *some* type function type function f (a : Type) = ... or, in Haskell, f :: Type -> Type (a function that takes a *concrete* type and creates another *concrete* type from it) that may have different constructor functions, just like **Maybe** or **List**? Can we define operator <$> for it? Something like

(<$>) :: (a -> b) -> f a -> f b

We want this operator, just like in case of **Maybe** and **List**, to take a function g :: a -> b and turn it into a function that is *lifted* into our type function f - h :: f a -> f b. Let's write out full signatures in our pseudocode to make sure we are clear on what means what. Operator <$> takes a function g(x:a):b (*takes a value of type* a *and returns a value of type* b) and turns it into a function h(x:f(a:Type)) :f(b:Type) (*takes a value of type* f a *and returns a value of type* f b).

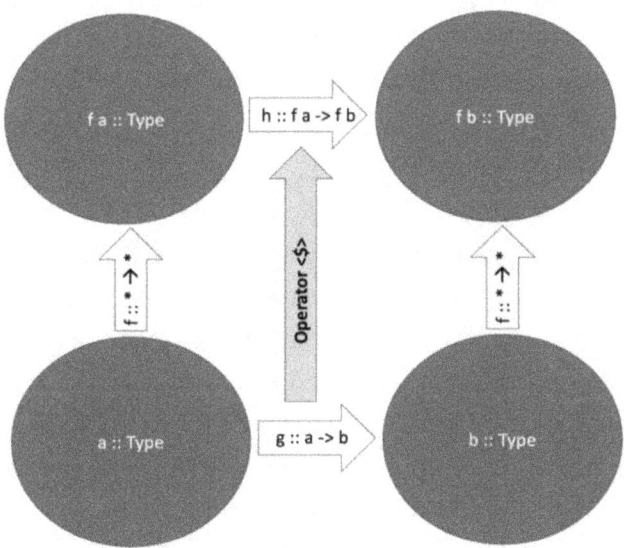

Figure 4-2. *Functor scheme*

CHAPTER 4   BASIC TYPECLASSES OR "SHOW ME A MONOID"

Congratulations! You just invented a **Functor**. A **Functor** in Haskell is merely any type function `f :: Type -> Type`, like **Maybe a** or list **[a]** that has an operator `<$>` defined, which "converts" or *lifts* functions between two concrete types – `g :: a -> b` – to functions between values of types constructed from "a" and "b" with the help of our type function f: `h :: f a -> f b`. In Haskell, **Functor** is a typeclass, the full definition of which will be given in the next chapter. In Category Theory, Functor is the foundational concept upon which it is built.

Why do we need Functors in Haskell? So that we can write `g <$> lst` and be sure our "simple" g function works on complicated values constructed with some type function – instead of resorting to verbose boilerplate nested `case` statements for each new type we define. Ah, yes, if you wrote operator `<$>` for lists from the exercise above, you should have noticed that's it's merely the `map` function. So **Functor** is our first very useful abstraction. If you define some type function `f::Type->Type` that takes some concrete type and creates another concrete type from it – similar to Maybe or List – always check if you can turn it into **Functor** by defining a `<$>` operator. This will make your code concise, easy to read, and a breeze to maintain. We will see many Functor examples further in this book.

Most of the confusion in Haskell stems from the fact that the concepts here are named almost exactly as they are in Category Theory, which is about the only actively developing area of mathematics nowadays, cutting edge of the most abstract of science. I mean, why not call this an *elevator* or *injector* or something less scary than a **Functor**?! But it is what it is: really simple elegant concept named so as to confuse us simple folk. However, for those who wish to be more technical and who have read the previous chapter, we will discuss Haskell Functors from the categorical point of view in the next chapter, along with their formal definition as a typeclass.

Conclusion

In this chapter we learned how to unleash the power of Haskell via typeclasses depending on one type parameter and constructed a couple of algebraic structures, such as Semigroup and Monoid, using this mechanism. At the end, by recalling functions we introduced when discussing **Maybe** type function, we "invented" a **Functor** concept – an extremely useful abstraction that makes Haskell code much more concise and maintainable. In the next chapter, we will define Functor formally and will start building upon this foundation.

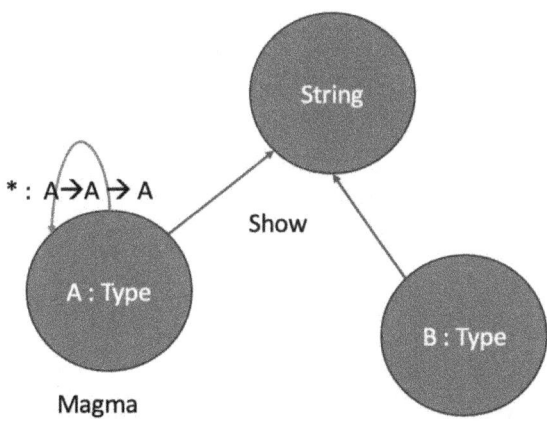

Figure 4-3. *Typeclasses add structure to your program category: Magma typeclass adds internal structure to a type (by providing endomorphism – binary operation A→A→A), Show typeclass adds structure to the whole program, by creating relations between your types (A and B in the picture) with String type*

When you design types of your problem domain – always look for structure inside types and between types. This structure can often be turned into existing typeclasses, such as a Functor, or you may get an idea for your own typeclass that will allow to abstract some of this structure and make your code more concise and efficient. Figure 4-3 simplistically illustrates this approach for designing programs.

CHAPTER 5

Functor, Bifunctor, and Applicative Functor Enter an Elevator ...

- "Thank you, Lambda!", you say enthusiastically as this beautiful hierarchy starts to make sense. "I really like the idea of using the same operators for different types using typeclasses. Now, about that contest ..."

- "Ah, yes, the contest. You'll have to conjure a card game using your spells and then hopefully win it. You created the deck already! Now you just have to finish the game! How about Blackjack?"

- "Wow, wow, hold on there, Lambda! I don't remember nearly enough to finish the game yet!", you start doubting yourself again.

- "Don't you worry, silly! We'll move slow. Let's first add the way to track players to our magical code now using a Functor and type functions?"

CHAPTER 5 FUNCTOR, BIFUNCTOR, AND APPLICATIVE FUNCTOR ENTER AN ELEVATOR ...

Functor Typeclass Definition

We will help our mage to further extend our card playing program further in this chapter. Before we do, we need to look at some theory in pictures.

Recall how we "invented" a **Functor** in the previous chapter when we needed a good way to apply functions between two concrete types $a \to b$ to values of type List a or Maybe a? To recap, in Haskell a **Functor** is merely any type function f :: Type -> Type, like **Maybe a** or list **[a]** that has an operator <$> defined, which "converts" or *lifts* functions between two concrete types, such as g :: a -> b, to functions between values of types constructed from a and b with the help of our type function f -- h :: f a -> f b (*see Figure 4-2*).

In Haskell, **Functor** is defined as a typeclass, which we will fully define and discuss below. Making your types an instance of **Functor** is a very good way to avoid writing a lot of boilerplate code. If you have some experience with imperative languages such as C++ or Java, it can be a good exercise to try to define a Functor for C++ templates or Java generics – which are as close as those can get to typeclasses. Once you spend an hour writing <$> for a List and a Maybe and then trying to generalize the approach, you will start appreciating the beauty and conciseness of Haskell a great deal.

Since both **Maybe a** and **[a]** are instances of **Functor**, you can write concise code such as (*2) <$> (Just 4) or (++ " hello") <$> ["world", "me"] instead of writing a *separate* function with nested case analysis statements each time. That's one of the reasons you can't evaluate Haskell programmers' performance in "lines of code" – if anything, we pride ourselves in having as few lines of code as possible to achieve a goal.

Strictly speaking, a **Functor** is not only a type function $f :: Type \to Type$ with the "lifting" operator <$> defined, but it also has some laws – just like a **Monoid**, which we discussed in the previous chapter. The laws are as follows:

1. (<$>) id == id: The identity function should remain an identity function even when lifted.

2. (<$>) (f · g) == (<$>) f · (<$>) g: Function composition must be respected – whether we lift a composed function f·g or lift f first, then g, and then compose them – the result should be the same.

These laws, just as the concept of Functor itself, come from Category Theory, where Functors are absolutely essential. In Category Theory Functor is defined roughly as mapping **F** between two categories **C** and **D**, which associates each object X of category **C** with an object **F(X)** in category **D** and each morphism in **C** f:X->Y to morphism in **D** F(f):F(X)->F(Y). Technically, in Haskell **F** corresponds to the type function such as **Maybe** or **List** (they provide object correspondence), while operator <$> takes care of morphism (function) correspondence. Since when writing Haskell programs we work in category **Hask**, then Haskell Functors are in fact *endofunctors* F:Hask->Hask. Functors are also called *structure-preserving maps* – think about why that is. Let's take a look at **Maybe** as a Functor. Let's say you wrote a very simple Haskell program where you use Strings, Ints, Doubles, and various functions between them. Strings, Ints, and Doubles are *objects* in your program category, functions between them morphisms, that provide *structure* for your program. However, your program is error-prone since you can't divide by 0, as the simplest example. You decide to make it safer and start working with Maybe Ints, Maybe Doubles, and Maybe Strings instead. In any imperative language you would have no choice but to rewrite lots of functions – absolute waste of time. In Haskell, since you *know* that **Maybe** is a **Functor**, you can be sure that simply by using operator <$> you will be able to use all of the "unlifted" functions with corresponding *Maybes* and not worry about breaking composition – it is *guaranteed* by math! Isn't it beautiful?

CHAPTER 5   FUNCTOR, BIFUNCTOR, AND APPLICATIVE FUNCTOR ENTER AN ELEVATOR ...

As a Haskell typeclass, **Functor** is defined with a bunch of additional useful functions with default implementation – again, just as we have seen with **Monoid**:

```
class Functor f where
    fmap :: (a -> b) -> f a -> f b
    (<$>) = fmap -- just an infix synonym
    (<$) :: a -> f b -> f a
    ($>) :: f a -> b -> f b
    (<&>) :: f a -> (a -> b) -> f b -- flipped <$>
    unzip :: f (a, b) -> (f a, f b)
```

So how would you use **Functor** abstraction in real life? If you know some type function is an instance of **Functor** – such as Maybe or List – you can use fmap or <$> to easily "lift" any a -> b function application to a value of such a type. When you design a *new* type function f :: Type -> Type that takes a type and creates another type out of it, think if you can make it an instance of **Functor** so that your code is concise and maintainable. It is trivially useful for all kinds of containers other than lists, but every time you see a data type in Haskell with type signature Type -> Type, it's worth checking if it is a **Functor** or whether it can be made one. This is also a good exercise to develop your intuition for other abstractions, such as **Applicative** and **Monad** that we will look at in the next sections. They are nothing scarier than a **Functor** – in fact, they *are* a **Functor** just with some more structure added!

FUNCTOR EXERCISE

Come up with a new type function Type -> Type that you could use in a real-world task. Can it be turned into a Functor? If yes, make it an instance of one by defining the fmap function for your type.

CHAPTER 5   FUNCTOR, BIFUNCTOR, AND APPLICATIVE FUNCTOR ENTER AN ELEVATOR ...

Three-Dimensional Vector Example

Let's work through a simplified example for the task above and then also give examples of usage for the additional Functor functions listed in the typeclass definition. Suppose we are developing a physics model where we need to store three-dimensional coordinates in a vector, but we may want to use **Double** or **Int** or even complex numbers as elements of our vector. We can define the following type for this purpose:

```
data Vector3 a = Vec3 a a a
```

Now, in real life there are *much* more efficient data structures for such a task, so please never model real physics with a data structure as above, but it helps work through such a simplified example to understand more complex types later on. Based on this definition, we can have Vector3 Double or Vector3 Int while defining polymorphic vector-related functions such as summation, dot product, etc.

Can we make our type an instance of **Functor**? It has the right type – Vector3 :: Type -> Type – as it takes a concrete type and makes another type out of it. To turn it into a Functor, we need to define the function fmap (or operator <$>) that will have the following type signature: fmap :: (a -> b) -> Vector3 a -> Vector3 b. Just by looking at the signature we can see such a function would be quite useful - we would definitely want to convert a Vector3 Int into Vector3 Double and since there is a function already that converts Int to Double (fromIntegral) – so if we have an fmap as above, we can simply lift it and not worry about writing a separate conversion function for each case!

Let's define our instance, quite trivial in this case – we are simply applying our function to each component of a vector and construct a new vector out of it:

```
instance Functor Vector3 where
    fmap f (Vec3 x y z) = Vec3 (f x) (f y) (f z)
```

CHAPTER 5   FUNCTOR, BIFUNCTOR, AND APPLICATIVE FUNCTOR ENTER AN ELEVATOR ...

VECTOR3 EXERCISE 1

Check if the Functor laws hold up for our Vector3 type. It is trivial for preserving identity, but a bit more tricky for preserving the function composition.

Now you can do many useful things with your vectors without the need to write separate functions! Scalar multiplication? (*2) <$> (Vec3 1 5 2) Conversion? fromIntegral <$> (Vec3 1 5 2) Neat! As we repeated many times, looking at and understanding type signatures is key to your productivity as a Haskell programmer. Let's compare the type signatures of the <$> operator and the operator of the "regular" function application $:

```
($)   ::            (a -> b) ->   a ->   b
(<$>) :: Functor f => (a -> b) -> f a -> f b
```

We can easily see from this that both of them apply one function f::a->b to in the first case a value of type a and in the second case a value of type f a. So, again – if you have a type that is a Functor (and there are lots of such types) – always remember about the operator <$>.

Now let's take a look at all the other functions that Functor typeclass gives us "for free" and think of some usage examples:

- (<&>) is just a flipped version of the (<$>), so we give a value first and the function second, e.g., [1,2,3] <&> (*2) – will multiply all the list elements by 2 – or, in the case of our Vector type, (Vec3 4 10 5) <&> (*2) and so on.

- ($>) and (<$) are flipped versions of each other and they allow us to change values inside a Functor type function to some constant, e.g.:

- Nothing $> 1 ➔ Nothing

- Vec3 50 17 29 $> 1 ➔ Vec3 1 1 1 – might be useful if we don't need the results of a previous computation that gives us our Vec3, but we will need a unit vector further on if the computation is successful. If it is not, we get Nothing, as mentioned above.

- unzip is a generalization of unzip for lists that you may be familiar with: unzip (Just ("Hello", "World")) ➔ (Just "Hello", Just "World") or, with our vectors, unzip (Just (Vec3 1 1 1, Vec3 1.0 1.0 1.0)) ➔ (Just (Vec3 1 1 1), Just (Vec3 1.0 1.0 1.0))

Tracking Players in Our Game

Now we know enough to design a useful type where we can store the game information for *any* game, not just Blackjack that Lambda suggested Curry to build. Every player in every game has some cards at any given point *and* some additional state – amount of money, current score, history, etc. We can design this type as follows:

data PlayerData a = PlayerData Deck a

If you recall, you should read this as `PlayerData (a : Type) = PlayerData( deck : Deck, state : a)` - **PlayerData a** is a type function that takes a type a and creates a new type that combines a **Deck**, which is simply a list of cards, and a variable of type a, which stores our state. Isn't this a great candidate for a **Functor**?! The type is right, and we will want to manipulate our state in many different ways during our game, so lifting the functions a -> a that change our state some way into our new Functor type `PlayerData a -> PlayerData a` will come very handy! Let's define a **Functor** instance:

```
instance Functor PlayerData where
  fmap :: (a -> b) -> PlayerData a -> PlayerData b
  fmap f (PlayerData deck x) = PlayerData deck (f x)
```

Now let's say we have a **PlayerData** with a simple **Int** as a state type (we will use more complex states down the line of course) that can represent an amount of money a player has, e.g., d1 = PlayerData [(Card Three Spades)] 10. This represents that a player has a Three of Spades in his hand and 10 dollars of money. What if our player wins 20 dollars? If we didn't have a **Functor** instance, we'd have to unpack our type, do pattern match, pack it back, etc. With **Functor** defined above, we simply write (+20) <$> d1 or, applying our knowledge of additional Functor functions listed in the previous section, even nicer-looking d1 <&> (+20) and we get a new data of PlayerData [(Card Three Spades)] 30! Beautiful, isn't it? This way we will design our state-manipulating functions separately without even thinking about PlayerData – and then will be able to automagically apply them to PlayerData, thanks to our Functor instance!

You Are Either Functor or a Bifunctor

By now you should have seen that there is nothing scary about a **Functor** save for a name. In this section, we will look at type functions from *several* type parameters and will easily invent something sounding even scarier than a Functor – a **Bifunctor**, which is again just another useful and natural abstraction.

So far, we have looked at the type functions with the type Type -> Type such as **Maybe a** or **[a]**. However, nothing prevents us from writing type functions with several type parameters – however many we want, just as with "real" functions. The simplest example of such a type is **Either**:

```
data Either a b = Left a | Right b
```

CHAPTER 5   FUNCTOR, BIFUNCTOR, AND APPLICATIVE FUNCTOR ENTER AN ELEVATOR ...

If we write down Either with a full type signature, it will be Either :: Type -> Type -> Type, which reads simply as "give me two concrete types and I'll make a new concrete type out of them." It has two data constructors and is very similar to **Maybe** with the key difference being that instead of Nothing as the second data constructor, here we can actually store some info along with it. As such, its most popular use is *also* for handling errors or failure, just as **Maybe**, but with additional information. Consider the safeDivide function we defined before:

```
safeDivide :: Double -> Double -> Maybe Double
safeDivide _ 0 = Nothing
safeDivide x y = Just (x / y)
```

In case someone divides by zero using our safeDivide, it won't break or throw an exception, it will simply return Nothing - so we can continue our safe computations. But what if we want to return some more information in case of errors or if we want to distinguish between errors? We can use **Either a b**! For instance:

```
verySafeDivide :: Double -> Double -> Either String Double
verySafeDivide _ 0 = Left "Please don't divide by zero!"
verySafeDivide _ 100 = Left "We don't want you to divide by 100 either for some inexplicable reason!"
verySafeDivide x y = Right (x / y)
```

This function takes two Doubles and returns a value of type Either String Double. From its definition the pattern of usage is pretty obvious - errors are marked with Left data constructor and "regular" results with Right. Pretty cool and more useful than **Maybe** in many more complex cases (of course we don't really need Either for safe arithmetic, but in many real-world cases it is *very* handy).

However, here we immediately run into an issue by asking the same question as in Chapter 2 - what if I want to square the result of our verySafeDivide operation? For **Maybe**, we have a **Functor** instance, so we can simply write `square <$> safeDivide 4 3` and it will work like a charm. But we cannot turn **Either a b** into a **Functor**, it has the wrong type for it! Functor expects `Type -> Type`, and Either has `Type -> Type -> Type`. So what do we do? Are we forced to write long and ugly "case" analysis functions to be able to use **Either**?

EITHER AS FUNCTOR

Think about how and if you can make Either an instance of Functor. Define a typeclass instance with your ideas.

Of course not, that is not the Haskell way! We can design another generic function and see if it can be abstracted away into some other typeclass similar to a **Functor**. What we need to apply `square` to `verySafeDivide` is a function that takes `Double -> Double` and makes it work with `Either String Double`, so that we can write something like `eitherLift square (verySafeDivide 5 2)` and it would work. Let's write it:

```
eitherLift :: (Double -> Double) -> Either String Double -> Either String Double
eitherLift f (Left err) = Left $ "Applied a function to an error message: " ++ err
eitherLift f (Right res) = Right (f res)
```

If we are giving an error (`Left` value) to `eitherLift`, it simply translates the error further. If we are giving a proper result (`Right` value), we apply the function and put it under another `Right` value. Clean and easy and *very* similar to what we did when we invented lifting for `Maybe Double`, right?

So how about we generalize it further, by looking at the above and squinting hard and deciding that what we really want is the function with the following type signature:

```
eitherMap :: (a -> c) -> (b -> d) -> (Either a b -> Either c d)
```

Just like our `maybeMap` took a function a `->` b and "lifted it" into **Maybe** by turning it into a function `Maybe  a  ->  Maybe  b`, here we take *two* functions – a `->` c and b `->` d – and lift them into **Either** by turning *both* of them into *one* function `Either  a  b  ->  Either  c  d`.

But why stop here? Let's take it a step further and make such a function not just for **Either**, but for *any* type function with type signature `Type -> Type -> Type`. This is very similar to what we did when inventing a **Functor**, only now we invented a **Bifunctor**:

```
class Bifunctor (bf:: Type -> Type -> Type) where
    bimap :: (a -> c) -> (b -> d) -> (bf a b -> bf c d)
```

So **Bifunctor** is a type function with type `Type -> Type -> Type`, i.e., the one that takes two type parameters and creates a new type, with function `bimap` defined, which lifts *two* functions `f :: a -> c` and `g :: b -> d` into our **Bifunctor** by turning them into *one* function `h :: bf a b -> bf c d`. Figure 5-1 should help you visualize the concept.

CHAPTER 5 FUNCTOR, BIFUNCTOR, AND APPLICATIVE FUNCTOR ENTER AN ELEVATOR ...

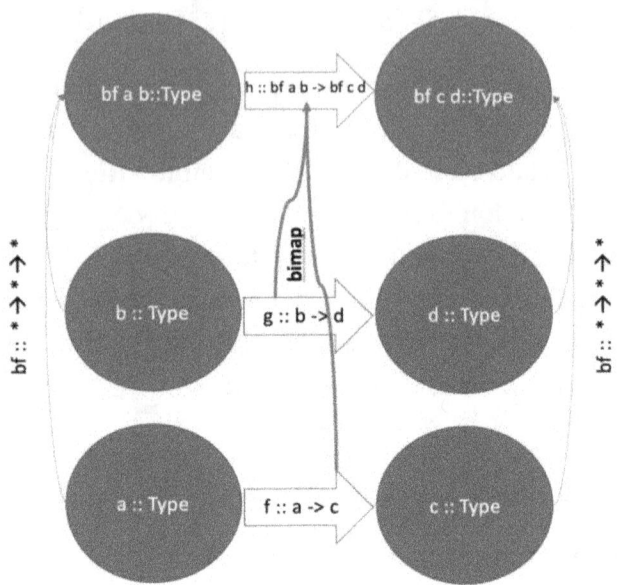

Figure 5-1. *Bifunctor scheme*

Do you notice a beautiful symmetry? `bf` takes two types and "lifts" them in a new type; `bimap` takes two functions and "lifts" them into a new one.

Making **Either** an instance of **Bifunctor** is very straightforward:

```
instance Bifunctor Either where
    bimap f g (Left  x) = Left  (f x)
    bimap f g (Right y) = Right (g y)
```

Again, in a couple of paragraphs of text, by solving a practical task of dealing with errors in our computations, we invented something with a very scary name, which is nothing more than a really cool and useful abstraction for type functions with two type parameters.

You should now be able to rewrite our pretty specific `eitherLift` function from above via general `bimap` function. Try doing it on your own before reading further.

To solve this task, you need to realize that we were lifting the `b -> d` function, in our case `Double -> Double`, while the `a -> c` one, in our case `String -> String`, was added implicitly in the body of the `Left` case. Realizing this, we can define `eitherLift` as follows:

```
eitherLift :: (Double -> Double) -> Either String Double ->
Either String Double
eitherLift = bimap (\err -> "Applied a function to an error message: " ++ err)
```

Please dwell on these two lines for some time until you fully understand what is happening here and you are fully convinced this definition is equivalent to the one given above. Strictly speaking, the function above will have a more generic type, `eitherLift :: (a -> b) -> Either String a -> Either String b`, so we will be able to lift not just functions of type `Double -> Double` with its help, but functions between any two types `a -> b`. Convince yourself why this is the case.

Can you invent a "Trifunctor"? Generalize it to "n-Functor" – a structure that combines an n-parameter type function and a corresponding "function lifter."

PLAYERDATA AS A BIFUNCTOR

We defined a Functor instance for our PlayerData type. However, values of this type store *two* values – a Deck and a state of type a. We may want to manipulate not just the state, but also the deck in our player data. Isn't it exactly what Bifunctor allows us to do? Try to define a Bifunctor instance for PlayerData. Is it possible?

CHAPTER 5   FUNCTOR, BIFUNCTOR, AND APPLICATIVE FUNCTOR ENTER AN ELEVATOR ...

We Need a Bigger Lift

So far, we've learned how to take "regular" functions of type a->b and "lift" them into various parametric types or type functions, such as Functors or Bifunctors. But what do we do with functions that take *two* parameters? For instance, what if instead of multiplying Just 4 by 2, which we can easily do by writing Just 4 <&> (*2) thanks to the **Functor** instance of **Maybe**, we wanted to multiply Just 4 by Just 2 (*for instance, if both arise as a result of some safe operation*)? We would need a function with the type Maybe Double -> Maybe Double -> Maybe Double, but we have no way to construct such a function easily! As another example, what if we have two lists of strings – ["hello", "good bye"] and ["world", "friend"] – and we want to stitch the strings in these lists together using (++) :: String -> String -> String? Here, the type of the function we need to construct is [String] -> [String] -> [String].

EXERCISE

Write these two functions – one that takes a function of type Double -> Double -> Double and turns it to Maybe Double -> Maybe Double -> Maybe Double and another that does the same for lists. The latter is more difficult; do you see why?

You should see how we would want to generalize it using the same approach as with **Functor** and **Bifunctor**. We need some operator, similar to <$> from **Functor**, that takes a function with the type a -> b -> c and lifts it into our functorial types by turning it into a function with the type f a -> f b -> f c. If you recall the picture for **Functor** from the previous chapter, it is a very simple extension of it to three types from two.

98

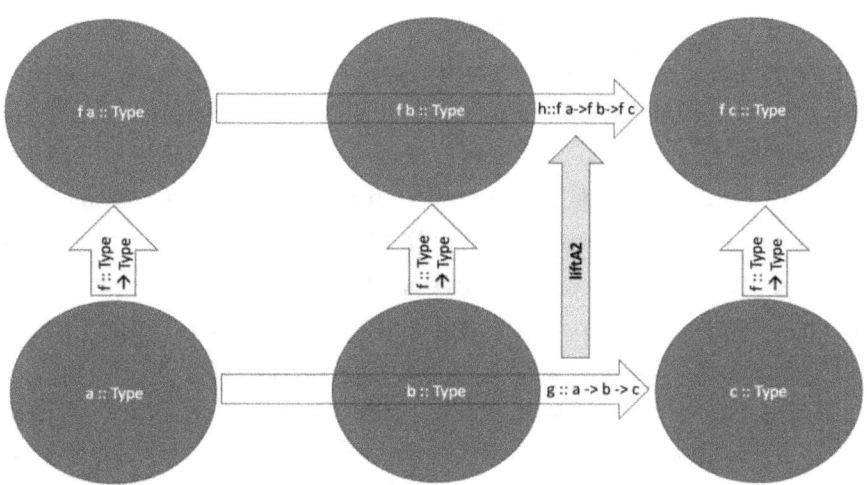

Figure 5-2. *Lifting a function of two parameters to a Functor: Applicative liftA2*

In Haskell, this is the (see Figure 5-2) operation `liftA2 :: (a -> b -> c) -> f a -> f b -> f c`, which is a part of the **Applicative** typeclass that we will fully define below. However, it turns out that such a simple, bordering-on-the-trivial extension of the **Functor** gives us more power than may seem at first glance. Before we discuss it, let's define the `liftA2` function for the **Maybe** type:

```
liftA2Maybe :: (a -> b -> c) -> Maybe a -> Maybe b -> Maybe c
-- if one of the arguments is nothing, we return Nothing:
liftA2Maybe _ Nothing _ = Nothing
liftA2Maybe _ _ Nothing = Nothing
-- otherwise, apply the function and put the result
inside maybe:
liftA2Maybe f (Just x) (Just y) = Just (f x y)
```

If you wrote it for `Maybe Doubles`, the general variant should be straightforward for you.

CHAPTER 5   FUNCTOR, BIFUNCTOR, AND APPLICATIVE FUNCTOR ENTER AN ELEVATOR ...

Now, what if rather than multiplying `Just 4` with `Just 2` as in the example above, we have a *function* inside our **Maybe** type? As you know, functions are first-class citizens in Haskell, so we may pass them around and return them as a result of other functions. So what if we have `Just (+2)` – a function that adds 2 to any other number we pass to it inside a **Maybe** – and we want to apply it to `Just 2`?

Or, looking at **List** instead of **Maybe**, what if we take a list of functions such as `[(+2), (*2)]` and we want to apply it to a list of numbers `[1,2,3]`? These two cases may seem unrelated at first glance, but you should already be able to identify type signatures quite well and be able to say that in the first case we have a value of type `Maybe (Int -> Int)` – i.e., a function `Int -> Int` *lifted* in our **Maybe** type – and we want to apply it to a value of type `Maybe Int`. In the second the list of functions is a value of type `List (Int->Int)`, written in Haskell as `[Int->Int]`, and we need to apply them all to a value of type `[Int]`.

A pattern emerges again – we have a type of functions *lifted* in a Functor and we need a way to apply them to a *value*, also lifted in *the same* Functor (Maybe in the first case, List in the second). To do this, we need another operator, which is also a part of the **Applicative** typeclass: `(<*>) :: f (a -> b) -> f a -> f b`. The signature tells us that we have a function `a -> b` lifted inside our Functor type function `f :: Type -> Type` and we need to somehow apply it to a value of type `f a` and get a value of type `f b` as a result.

On one hand, what it does is not really connected to `liftA2` function we defined above, but it turns out you can express one in terms of the other. This hints that two abstractions – "lift a function of two parameters into a type function" and "apply a lifted function of one parameter to a lifted value" – are very closely connected. Turns out, not only are they connected, but `<*>` operator provides an ability to chain Applicative computations or "lift" a function of any number of parameters into our Functor!

> **EXERCISE**
>
> Before reading further, try to define `liftA2` via operator `<*>` and vice versa.

Now let us compare main Functor's and Applicative's operators by staring attentively at their type signatures:

```
($)   ::                    (a -> b) ->   a ->   b
(<$>) :: Functor f =>       (a -> b) -> f a -> f b
(<*>) :: Applicative f => f (a -> b) -> f a -> f b
```

Do you again see this beautiful symmetry and rhymes? First, "regular" or *pure* function application. Then, application of the pure function to the lifted value. Then, application of the lifted function to the lifted value. Here is a very simple example on how **Applicative** makes your code much easier to read and maintain. Let's say you have the following expanded Person data type:

```
data Person = Person {
     name :: String,
     age :: Int,
     children :: Int
}
```

Then let's assume you have *safe* computations (so ones that return corresponding *Maybes*) that give you the name of the person, their age, and number of children:

```
fName     :: Maybe String
fAge      :: Maybe Int
fChildren :: Maybe Int
```

This is a quite common scenario if you are getting this data from the user via keyboard input, reading a database, etc. The question is, how do you construct a new Person instance from these three functions? The type of the Person constructor is the following: `Person :: String -> Int -> Int -> Person`. It expects pure values. You have lifted values. Anything from the Functor typeclass doesn't help you in the slightest – if you do `Person <$> fName`, which is equivalent to `fmap Person fName`, here is what you'll get:

`fmap::(a -> b)->f a ->f b (Person:: String -> Int -> Int -> Person fName::Maybe String)` ➔

fmap expects `(a -> b)` and `f a` as arguments and we give it `(String -> (Int -> Int -> Person))`, which instantiates b as `Int -> Int -> Person`, and `Maybe String`, which instantiates a as `String`. So the result of such fmap application produces `f b`, or in this case `Maybe (Int -> Int -> Person)` – and we are stuck! We have no way to apply a function lifted into **Maybe** to anything else if we are limited to **Functor**! What do we do?! If we are forced to construct the new **Person** value by hand, we will be left with the following ugliness:

```
newPerson (Nothing _ _) = Nothing
newPerson (_ Nothing _) = Nothing
newPerson (_ _ Nothing) = Nothing
newPerson (Just s) (Just a) (Just c) = Person s a c
```

Fortunately, I am sure by now you are able to quickly spot type signature similarities, and `Maybe (Int -> Int -> Person)` tells you that it's very similar to `(<*>) :: f (a -> b) -> f a -> f b` from **Applicative**, which we introduced above! If you spend five minutes and unwind it further, you will quickly see that the alternative way to define newPerson knowing that **Maybe** is also an instance of **Applicative** is as follows:

```
newPerson fName fAge fChildren = Person <$> fName <*> fAge <*> fChildren
```

Much, much more beautiful! This way you can chain any number of functorial computations or "lift" functions of any number of parameters into your Applicative Functors.

Before we see the full **Applicative** typeclass definition and discuss most of its functions, let us think whether our **PlayerData** type – that we already made an instance of **Functor** – can be made an instance of **Applicative** and how it will help us in winning the card playing program competition. **Applicative** allows us, in essence, to "lift" pure functions of any number of parameters to our **Functor**, in this case, **PlayerData**. Can we imagine situations where we need to manipulate several PlayerData states at once? Recall how it is defined:

```
data PlayerData a = PlayerData Deck a
```

Here, Deck contains the cards our player currently has, while the variable of type a contains any state relevant for the game. For example, for a simple game of Blackjack, it can simply be the amount of money a player has, while for Poker it can include info on bids a player made during various bidding rounds and so on. For games such as Poker, we will *definitely* want to manipulate several states at once – e.g., an algorithm will want to know the moves previous players made before deciding on which move should the current player make. For Blackjack it's not that important, but at the very least we may want to compare the amount of money (or score) each player has to print out leadership boards and so on. This is enough of a justification to try to define an **Applicative** instance for **PlayerData**.

To do that, we need to define two functions – pure and one of <*> or liftA2. Let's start with the latter and then come back to pure, which is new. Making usual type instantiations, we get the following type signature for liftA2 in the case of PlayerData:

```
liftA2PD :: (a->b->c) -> PlayerData a -> PlayerData b ->
PlayerData c
```

> **TRY TO DEFINE IT BEFORE READING FURTHER**
>
> You are given the function of two variables, x::a and y::b, and you need to apply it to the two values of types PlayerData a and PlayerData b correspondingly. As usual with Haskell, once you figure out the types, actual functions are quite straightforward (notice the pattern match again):
>
> ```
> liftA2PD func (PlayerData deck1 sa) (PlayerData deck2 sb) =
> PlayerData ??? (func sa sb)
> ```

As you can see, we have a problem. We may have two different decks in the first PlayerData and in the second. If that's the case, we need to somehow combine them, but there is no logical way to do it without breaking the **Applicative** laws (see below in the typeclass definition). If you attempted to make PlayerData an instance of **Bifunctor** previously, you by now realize that it is impossible as well – Bifunctor requires the Type->Type->Type signature and our PlayerData has only Type->Type, but that is really unfortunate, since we *know* that **PlayerData** *is* a **Bifunctor**! This hints that our data type design is not really optimal – there are some very useful abstractions we could use, but with the way we defined our types, we can't. This is another very cool feature of Haskell – if you carefully think about your types and the structure between them, even that exercise will force you to design *better* types! We will improve our PlayerData design down the line, and for now let's come back to **Applicative**.

Action! Apply! Cut!

If you read the docs on **Applicative** in the Haskell base library, you probably noticed there is a lot of talk about *actions* and *computations*. It seems that it somehow helps sequence them efficiently (one example with

CHAPTER 5   FUNCTOR, BIFUNCTOR, AND APPLICATIVE FUNCTOR ENTER AN ELEVATOR ...

Person we have seen above). To be able to see what the fuss is all about on a simpler example than Parsec parser library (which is amazing but out of scope of this book – however, if you want to really appreciate the power of **Applicative**, that is probably the best library to study; we will also have another look at **Applicative** when discussing data structures and **Traversable** typeclass), let us invent another interesting type function of one parameter f :: Type -> Type, as you are probably tired of **Maybe** and **List** by this point.

As you may recall, the functional way to solve a problem is to compose different functions together, making sure the types "fit." So what if we want to record some information as we are executing our functions and then read it at the end of our program? Haskell is a pure language, and even though we can work with mutable variables, it should be our last resort for a lot of reasons, which we will discuss later on. But if we do not use mutable state, how do we record this information?! Our functions are pure mathematical functions, they take one value and give another value, and there are no side effects. Well, this hints that we need to somehow pass our information along the whole chain of our functions. For an easy but interesting example, let us count the number of function calls in our program.

We have a bunch of pure functions in our program with signatures such as a -> b, b -> c, c -> d, etc. that compose very well. We need some way to amend our types to also somehow pass the counter. For example, in a similar vein with our "bad" PlayerData:

```
data Countable a = Countable Int a deriving (Show)
```

Now if we have some way to turn our a -> b functions into Countable a -> Countable b, we should be able to achieve what we want. What operator do we have that takes a -> b and turns it into f a -> f b? Of course this is simply fmap from the **Functor** typeclass! Let's make our Countable type an instance of one:

```
instance Functor Countable where
```

105

CHAPTER 5 FUNCTOR, BIFUNCTOR, AND APPLICATIVE FUNCTOR ENTER AN ELEVATOR ...

```
fmap f (Countable c x) = Countable (c+1) (f x)
(<$) x (Countable c y) = Countable c x
-- ^^^ needed to make further Applicative instance work
correctly
```

> **EXERCISE**
>
> Can you identify the problem with the Functor instance definition for Countable above? Do you have an idea how to fix it?

On the surface, this definition does what we want. You can try something like this in REPL to make sure:

```
ghci> let x = Countable 0 10
ghci> show x
"Countable 0 10"
ghci> x <&> (*2)
Countable 1 20
```

So far, so good. However, if you recall, every Functor instance must satisfy certain laws. First, let's check for identity: `fmap id == id`. Does it hold? Let's check:

```
fmap id (Countable c x) = Countable (c+1) (id x) =
Countable (c+1) x
```

Oh-oh. While fmapping identity keeps the value of the function the same, it still increases our counter by 1, so the values on the right and on the left are *not* identical! There is a way around it. Quite possibly that from the pure mathematical point of view this is incorrect and at least reeks of cheating, but we can argue that from the point of view of our program execution we only care about values, not the counter. That is, we can say that Countable 15 "Hello" is actually equal to the Countable 394 "Hello", since we only care that "Hello" == "Hello". The counter is only there so that at the end of our program's execution we can check how many steps it took. Philosophically, we can say that our values in Countable live in our "program universe," while the counter is outside of it. Regardless of what you think about this train of thought, we indeed *can* redefine the equality function for our **Countable** type and this way force it to be conformant to the **Functor** laws as follows:

```
instance Eq a => Eq (Countable a) where
     (==) (Countable _ x) (Countable _ y) = x == y
```

Here we are making **Countable a** an instance of the **Eq** typeclass that deals with equality by redefining the (==) operator and making it ignore the counter value. As you can by now see, via typeclass manipulation you can create pretty much anything in Haskell. The second Functor law is composition, which says fmap (f . g) == fmap f . fmap g. This presents another problem – it is quite obvious that the counter will increase by 2 on the right and only by 1 on the left, so in effect we will be counting not the *pure function* applications, but rather applications of fmap (*even though with the new definition of equality of Countables the law formally holds*). Let's move to the full **Applicative** typeclass definition and further development of **Countable** type.

Applicative Typeclass

The full **Applicative** typeclass has the following functions:

```
class Functor f => Applicative f where
     pure :: a -> f a
```

CHAPTER 5 FUNCTOR, BIFUNCTOR, AND APPLICATIVE FUNCTOR ENTER AN ELEVATOR ...

```
(<*>) :: f (a -> b) -> f a -> f b
(<*>) = liftA2 id
liftA2 :: (a -> b -> c) -> f a -> f b -> f c
liftA2 f x = (<*>) (fmap f x)
(*>) :: f a -> f b -> f b
(<*) :: f a -> f b -> f a
```

To define an **Applicative** instance for your type, you have to first define a **Functor** instance, then define pure that "lifts" a pure value into your **Functor**, and then define one of <*> or liftA2. Before we look at how you can use additional functions, let's make **Countable** instance of **Applicative** and then check if it satisfies the Applicative laws. We will define both liftA2 and <*>, as it will make checking the laws easier:

```
instance Applicative Countable where
    pure x = Countable 0 x
    liftA2 func (Countable c1 x) (Countable c2 y) = Countable
    (c1 + c2) (func x y)
    (<*>) (Countable c1 func) (Countable c2 x) = Countable
    (c1 + c2) (func x)
```

By now you should be able to write such definitions without any issues. Here for pure we simply "lift" our value into Countable and initialize the counter to 0, and when we lift a function of two parameters into our Functor, we make sure the counters are added together – both very logical assumptions about the behavior we want. Let's check the laws:

- Identity: pure id <*> v = v ➔ pure id = Countable 0 id, (Countable 0 id) <*> (Countable c x) = Countable (0 + c) (id x) = Countable c x == v. Holds even without redefining equality!

- Composition: pure (.) <*> u <*> v <*> w = u <*> (v <*> w). This is a bit bulky to expand, but worth doing as an exercise: (.) :: (b -> c) -> (a -> b) -> (a -> c), then we have

(Countable 0 .) <*> Countable c1 vu <*> Countable c2 vv <*> Countable c3 vw = Countable c1 vu <*> (Countable c2 vv <*> Countable c3 vw). I leave the full expansion as an exercise, but it is very easy to see that the order of function compositions holds, while the counters will be correctly accounted for as well, since 0+c1+c2+c3 = c1+c2+c3. Again, no need for redefined equality.

- Homomorphism: pure f <*> pure x = pure (f x). We have (Countable 0 f) <*> (Countable 0 x) = Countable 0 (f x) == pure (f x). Holds!

- Interchange: u <*> pure y = pure ($ y) <*> u. This can be a bit confusing at first, but again, the trick is to carefully expand the function definitions. We have on the left: (Countable c1 vu) <*> (Countable 0 y) = Countable c1 (vu y). On the right: (Countable 0 ($ y)) <*> (Countable c1 vu) = Countable c1 (($ y) vu) = Countable c1 (vu y) – the last expansion requires some thinking about the function application operator: ($) :: (a -> b) -> a -> b, defined as ($) f x = f x. Convince yourself that ($ y) func == func y and you will see that interchange holds for our Countable as well.

Great! By now we have defined the **Applicative** instance for our **Countable** type and even checked that all four laws hold even without the equality redefinition trick that we had to resort to, to make it work for the **Functor** instance. Let's look at a small example program to illustrate usage:

```
printCosts st costs curr = "Your costs are " ++ show ((length st) * costs) ++ " " ++ curr
```

```
stringComp = Countable 3 "Hello World"
costComp = Countable 5 10

printCostsA = printCosts <$> stringComp <*> costComp <*>
(pure "USD")
```

Here, we have a pure function `printCosts` that takes a string, cost per symbol, and currency string and returns how much you owe. Then we have two computations (that we assume were a result of a bunch of other computations) represented by `stringComp` and `costComp`, the first done in three steps and the second in five steps. Then `printCostsA` sequences our computations to calculate the final result. You can try it in REPL:

```
ghci> printCosts "Hello" 10 "USD"
"Your costs are 50 USD"
ghci> printCostsA
Countable 9 "Your costs are 110 USD"
```

As you can see, `printCostsA` correctly sums up the number of steps in each computation and adds 1 for the `printCosts` application. Convenience operators `*>` and `<*` let you sequence computations and ignore previous (or future) computation results, while keeping the effects! This is a very cool functionality, very useful for parsers or other effectful computations you can come up with, but our simplistic **Countable** case illustrates it very well:

```
ghci> stringComp *> costComp
Countable 8 10
ghci> stringComp <* costComp
Countable 8 "Hello World"
```

As you can see, in both cases effects are being taken care of (the counter from both computations is added), while in the first case we are left with `costComp` value (10) and in the second with `stringComp` ("Hello World").

In most cases you will be using **Applicative** functions from existing libraries, but if you see that you have some effectful computations in your program that can benefit from being made **Applicative**, now you will be able to!

Conclusion

In this chapter we looked at **Functor**, **Bifunctor**, and **Applicative** definitions on a variety of specific examples and found out that our **PlayerData** type design is not optimal. This chapter should lay a very solid foundation for your ability to think about your program as a category inhabited by your types and about typeclasses as entities capturing the structure of your category – relationships between types. With some additional exercises, you will be able to quickly identify various abstractions in your problem domain and write programs that are concise, beautiful, and efficient. In the next chapter, we will look at the next – extremely important for Haskell – step in the *Functor–Applicative–Monad* hierarchy: Monads.

CHAPTER 6

O, Monad, Help Me Compose!

By now, you should have a basic intuition on how various Haskell abstractions arise. We have concrete types, such as **Int** and **Char**; we have "type functions" that create new types from other types, such as **Maybe a** and **List a**. We have typeclasses that capture internal types' structure as well as relations between them. In the real-world application, that is where you should always start: design your types, then try to capture relations between them, and see whether they correspond to some of the known abstractions, such as **Monoid**, **Functor**, or **Applicative**. Once you have, the only thing left is decomposing the problem into smaller pieces and solving it by composing functions together, just like in the simplified picture from Chapter 1 (Figure 6-1).

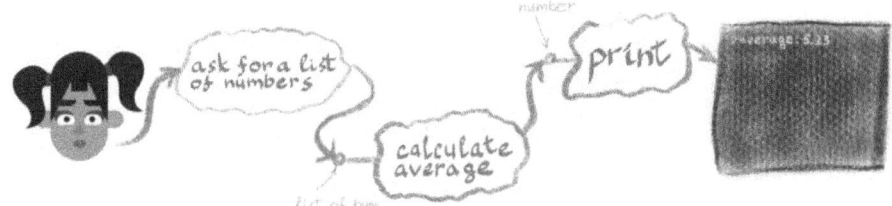

Figure 6-1. *Decomposing the problem functionally*

CHAPTER 6 O, MONAD, HELP ME COMPOSE!

All of the abstractions we have and will still look at in this book serve just this purpose: they help make our functions compose in a beautiful, concise, safe, and maintainable fashion.

In the previous chapter, we defined and examined in detail the **Functor** typeclass, its direct "child" – **Applicative** – and another naturally arising abstraction, **Bifunctor**. Functors allow you to apply pure functions to "lifted" values. Applicative Functors give any kind of computations with effects a very natural interface, as well as an ability to "lift" pure functions of many parameters into your Functors. Monads are nothing more and nothing less than a very natural extension of **Applicative Functors** that gives an extremely powerful framework to structure real-world applications via so called *Monad transformer stacks*. We will look at the foundation and many practical examples of them in the second part of the book. For now, let's focus on **Monads** themselves.

WARMUP EXERCISE

Come up with an example of a type that is a Functor, but not an Applicative (besides the PlayerData from the last chapter).

By now we have learned to identify abstractions, read type signatures, and define typeclass instances, so let us start with the main new function that the **Monad** typeclass adds compared with **Applicative**:

```
class Applicative m => Monad (m :: Type -> Type) where
    (>>=) :: m a -> (a -> m b) -> m b
```

CHAPTER 6   O, MONAD, HELP ME COMPOSE!

Let's look at the version of the >>= operator with its arguments flipped in the context of the Functor-Applicative-Monad hierarchy and think about similarities and differences. We have also added parentheses on the right side to drive a point about "functions conversion" across (see Figure 6-2 for visualization):

```
($)   ::                       (a -> b  ) -> (  a ->   b)
(<$>) :: Functor f =>          (a -> b  ) -> (f a -> f b)
(<*>) :: Applicative f => f    (a -> b  ) -> (f a -> f b)
(=<<) :: Monad f =>            (a -> f b) -> (f a -> f b)
```

Let's forget about the pure function application in the first line and consider the other three. It is easy to notice that all three operators <$>, <*>, and =<< convert some slightly different function in each case to a function between lifted values g :: f a -> f b. Functor works on pure functions a->b, Applicative works on lifted functions f (a -> b), and Monad works on the functions of the new curious kind: a->f b.

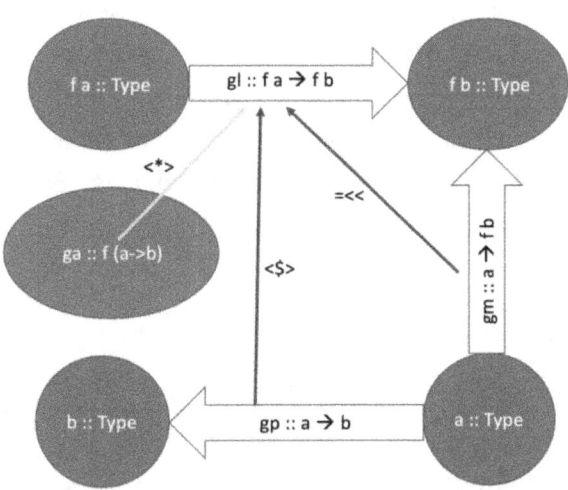

Figure 6-2. *Functor, Applicative, and Monad all convert different kinds of functions into one – a function between two lifted values* $fa \rightarrow fb$

115

CHAPTER 6   O, MONAD, HELP ME COMPOSE!

Intuitively, Monads are different from Applicative in a sense that an input value to a monadic computation can influence both its result *and* effects, while for Applicative effects are always there in any chained computation and we can only influence the results – and this is not what you always want. Another key point about monads is that *there is no default way to "escape" them*: our monadic computations always return values lifted in the monad, and even though the >>= operator produces a pure value, it can only be chained with another computation that returns a monadic result. This is one of the key reasons why all of the IO in Haskell is done inside a monad – so that the effects do not "escape" and pollute our pure functional world, but more on that below. For now, let's dig into Monads.

If you, like me in the beginning, read a lot of confusing "monad tutorials," I encourage you to forget them all and just think about **Monads** as what they are – another extremely useful abstraction building on the foundation of **Functor** and **Applicative**. As any abstraction, it can be applied to a bunch of quite different entities, and we will review the most important of them in this chapter.

Let's start by building upon the **Countable** type that we defined in the previous chapter and that counts the number of times we applied pure functions via the <$> operator from **Functor** typeclass. Since it already has instances of both **Functor** and **Applicative**, let's make it an instance of **Monad**, too:

```
instance Monad Countable where
    (>>=) (Countable c1 x) f = let (Countable c2 y) = f x
                               in Countable (c1 + c2) y
```

Dwell on this definition a bit as it is a bit tricky. What we are doing here is applying our monadic function f :: a -> Countable a to the value we extract from the Countable argument, pattern-match the result to get the new counter c2, and then add counters in the final result. What additional flexibility does this give us compared with **Functor** and **Applicative**? As mentioned above, it is the ability to manipulate the effects (in this case our counter) directly. Whereas previously our counter counted only one thing – number of times operator <$> was used – now we can do anything we want with it (code from the last chapter for your reference):

```
stringComp = Countable 3 "Hello World"
```

```
-- new "expensive" computation that manipulates the counter
directly:
veryCostlyFunction :: String -> Countable Int
veryCostlyFunction s = Countable 100 (length s)
```

```
-- and then in repl:
ghci> stringComp >>= veryCostlyFunction
Countable 103 11
```

Do you understand what is happening here? veryCostlyFunction expects a String as an input, calculates its length, and adds 100 as counter. Operator >>= in our **Countable** monad extracts the String value "Hello World" from the stringComp and executes veryCostlyFunction while adding both counters! This way, we get 11 – length of "Hello World" – as a result and 103, sum of our counters, as the effect. As you can see, now we can manipulate the effects directly in our monadic computations, while at the same time they are threaded "automagically" thanks to the >>= operator.

Can We Play Cards in This State?

All right, it is time to help our mage Curry to take another step toward winning the card playing competition! If you recall, in the last chapter we suffered a little setback – our **PlayerData** type, defined as PlayerData a = PlayerData Deck a, turned out to be not a very well-behaved one.

CHAPTER 6   O, MONAD, HELP ME COMPOSE!

We could make it an instance of a **Functor**, but not **Applicative**, and not even a **Bifunctor** – even though we *know* it is one.

But let us now take a step back and follow our own advice: before writing actual code for the Blackjack game, let's think about the problem domain and *design the types, including function types*. We already have the types for the most important thing in the game – actual cards. But what are the other entities that we need to model? Also keep in mind that ideally we want to design the types abstract enough so that we can model *any* card game, not just Blackjack. In any game we have players, sometimes of different types (e.g., in Blackjack there are normal players and a dealer). Players have some "global" game state – such as a score, amount of money, etc. – as well as "local" game state, which is used to track what is happening during each round, such as different types of cards (visible, hidden, etc.), bets, moves made, etc.

Let's try modeling these as generally as possible, but starting with a simplified Blackjack game. The rules will be as follows:

- The dealer deals two cards to each player and one card to themselves.

- Players make bets (we will not allow any splits and others to simplify the game).

- Players take cards until they say "stop" or go bust (over 21 total score).

- The dealer takes cards until they hit 17 or go bust.

- The score is calculated as 10 for J, Q, and K, either 1 or 11 for Ace, and at face value for the other cards.

- If the player is closer to 21 than the dealer, they get paid their bet; otherwise, they lose it.

- If the player gets 21 from the initial dealing, they win 2× their bet.

CHAPTER 6   O, MONAD, HELP ME COMPOSE!

Let's start with the dealer, as they have no real choices in this game but to deal cards to players, deal one card to themselves, wait for the players to bet, and then take cards until they hit 17 or go "bust" (i.e., over 21).

The "global" game state for the dealer is equal to their deck (we think that their money is unlimited). The "local" round state adds one visible card in the beginning and then repeatedly takes cards as described above. This looks very easy, and we can model everything via two decks - one from which we deal and the other one with the current dealer's cards.

For the player, we need their current money as a "global" state and then their cards and their bet as a "local" round state.

Before we define the final types for these entities, let us think about a typical action we'll have to take repeatedly in the game: dealing the cards. To do this, we need to take the dealer's deck :: Deck (if you recall, the Deck is just a list of Cards), take n cards from it and create another Deck from them, and change the dealer's Deck correspondingly.

If you have been following our gradual buildup of the **Hask** world closely, you will immediately notice two problems: Haskell's values are immutable - which in effect means every function returns a *new* value - and there is no global state.[1] This means that we have to somehow thread the state (dealer's deck) throughout our computations - just as we did with the Countable's counter above. So our function should have something like the following type: dealCards :: Int -> Deck -> (Deck, Deck). We give it a dealer's deck, take n cards from it, and return the new (updated) dealer's deck and these cards separately (so that we can pass them to the user and so on). Luckily, Haskell has very powerful list manipulation functions in the **Data.List** package, so to implement this function we can simply write

---

[1] Strictly speaking, of course there are mutable variables (and thus global state is possible) in Haskell, and we will discuss them in this book. However, their usage is discouraged unless you absolutely have to - an immutable, pure functional world provides mathematical guarantees for safety, program correctness, etc. that are simply too good to overlook. So whenever you can remain pure, do so.

CHAPTER 6   O, MONAD, HELP ME COMPOSE!

```
dealCards :: Int -> Deck -> (Deck, Deck)
dealCards n deck = let newDeck = drop n deck
                       plDeck  = take n deck
                   in (newDeck, plDeck)
```

By now reading this function should be very easy for you – it "returns" (newDeck, plDeck) and they are being calculated by the drop and take n, respectively. You can try it by loading the Cards code into ghci:

```
ghci> let x = dealCards 3 fullDeck
ghci> x
([5♣,6♣,7♣,8♣,9♣,10♣,J♣,Q♣,K♣,A♣,2♠,3♠,4♠,5♠,6♠,7♠,8♠,
9♠,10♠,J♠,Q♠,K♠,A♠,2♦,3♦,4♦,5♦,6♦,7♦,8♦,9♦,10♦,J♦,Q♦,K♦,
A♦,2♥,3♥,4♥,5♥,6♥,7♥,8♥,9♥,10♥,J♥,Q♥,K♥,A♥],[2♣,3♣,4♣])
```

Great, it does what we want. However, what if we want to deal some more cards afterward? Or what if the players played their hands and we need to return their cards to the dealer's deck? Are we forced to design all our functions to work with clumsy pairs (Deck, *something*) – extract state, extract *something*, manipulate, pack the results back ... Ugh, even thinking about this give me shrieks! What if our global state is more than just a Deck, but lots of other parameters?

Let us add one more function to illustrate the situation better – the one that counts the current number of cards in the "global" (dealer's) deck. Since we need to thread our global state from computation to computation, we need to pass it explicitly again, so the function becomes

```
countCards :: Deck -> (Int, Deck)
countCards deck = (length deck, deck)
```

If you squint real hard, you should notice the similarity between the two functions we just defined and the function that is part of the >>= monadic operator: (a -> m b). For the function dealCards, a = Int and m b = (Deck -> (Deck, Deck)), so b = Deck, and for the function countCards there's no a and its type is just m b = Deck -> (Int, Deck), so b = Int.

What follows is somewhat advanced: we are going to design what is in fact a so-called State Monad from the ground up. This construction introduces several new topics, such as function types, a new Haskell keyword, and then the definition of the Functor, Applicative, and Monad instances for these new types. This will teach you how to build *new* Monad types yourself – but it is not really necessary to become a proficient practical Haskell programmer: you can simply use the interfaces for the Monads that have already been defined for you (we will discuss them in detail in further sections), and they will be enough for 95% of cases. So it is okay to skim or skip the next section on the first reading if the material seems too difficult for you and revisit it later once you become fluent in using the standard Reader–Writer–State monads.

If you recall our translation of what Haskell's notation of m a means, we can think of m as a *type function* that takes a concrete type a and creates a new type m a out of it. We've looked at many examples of such type functions by now – **Maybe, List, Countable, and PlayerData** – but none of them had a -> in their definition. This is nothing to be afraid of – as we also discussed, *functions* are first-class citizens of our **Hask** world, which means that their types are just regular types as well. In the case of the two functions we defined above, the definition for our m a is Deck -> (a, Deck). In other words, this is the type of functions, or **computations**, that take a **Deck** (in our case "global," or dealer's, deck) and return a pair of this same Deck (potentially altered – hence, it's a "global state") as well as a result of our computation of type a.

CHAPTER 6   O, MONAD, HELP ME COMPOSE!

Let us define this type formally:

`newtype DeckM a = DeckM { runDeckM :: Deck -> (a, Deck)}`

If this looks scary, let's unwrap this definition step by step. First, a new keyword: newtype. Basically, it introduces an, uhm, new type, much like the data keyword we've used before. However, there are some subtle differences: First of all, you can only use the newtype with types of one constructor (so no sum types such as **Maybe** or **Either** can be defined with newtype). Second, it makes your code a bit more efficient, because newtype completely erases any indirection in the way the data is stored at runtime. Third, while providing this efficiency, it also gives us an opportunity to hide the internal representation of our **DeckM** type, which makes interfaces cleaner and easier to use and maintain.[2] For all practical purposes, it's the same as the data type, but more efficient at runtime for single-constructor types.

On the right side of the **DeckM** definition, we have a simple record with only one field: runDeckM, which is in fact the function, or computation, with the type signature we recovered above: Deck -> (a, Deck) that – as we will soon see – will allow us to neatly thread the global Deck through all our computations without referring to it explicitly. Why are we putting our computation inside a record instead of simply writing type DeckM = Deck -> (a, Deck)? We could, but then we couldn't hide the internal representation of our type – and this is not a good practice in this case.

Okay, now that you understand the definition of DeckM, it's time to roll up the sleeves and make it a Monad.

---

[2] There are much deeper differences between newtype and data, having to do with how both keywords represent the "bottom" values, but this is beyond the level of discussion in this book and also not very relevant from the practical point of view.

CHAPTER 6   O, MONAD, HELP ME COMPOSE!

ADVANCED EXERCISE

Try to define Functor, Applicative, and Monad instances for the DeckM type we just described. Remember the laws for all these typeclasses – first thing that comes to mind might not be correct!

To make any type function m a (or, in our pseudocode from the first chapters, m(a:Type)) an instance of the **Monad**, first, we have to define instances of **Functor** and **Applicative**, since they form a hierarchy (arguably the most important in Haskell). We have made many type instances of **Functor** and **Applicative** already, and we also just made **Countable** a **Monad** – so you have a starting point if you want to try to do it yourself (which we encourage you to).

Let's do it step by step and you can compare with your solution. First, the **Functor**. To make **DeckM** an instance of **Functor**, we have to write a definition for fmap :: (a -> b) -> DeckM a -> DeckM b – a function that "lifts" pure function a -> b into a function that converts a *computation* DeckM a into DeckM b. This is the tricky part – **Functor** for types such as **Maybe** is trivial, but we have an actual function inside our type, and if we unpack the DeckM type, we will realize that this fmap has to convert computation of type Deck -> (a, Deck) into Deck -> (b, Deck), which can be difficult at the first attempt, but once you do it a couple of times you'll get the gist of it. Here's the definition:

```
instance Functor DeckM where
    fmap :: (a -> b) -> DeckM a -> DeckM b
    fmap f (DeckM {runDeckM = runDeckOld}) = DeckM { runDeckM =
    \deck -> let (x, d) = runDeckOld deck in (f x, d) }
```

Ouch! That's a mouthful. Let's again unwind it carefully. (DeckM {runDeckM = runDeckOld}) is a pattern match for our DeckM a value that fmap expects along with the function f::a->b. What it does is it lets us refer

CHAPTER 6   O, MONAD, HELP ME COMPOSE!

to the "hidden" `runDeckM` computation inside of our `DeckM a` by name `runDeckOld`. Thus, `runDeckOld` has the following type signature: `Deck -> (a, Deck)`.

On the right, we have to provide a `DeckM b` value, which is a `DeckM` record that contains a new `runDeckM` field – computation with type signature `Deck -> (b, Deck)` where the value of type b has to be somehow produced from the value of type a in `runDeckOld` with the help of the function f. If your head is spinning, just re-read all of the above and carefully study the type signatures. Types are *always* the key in Haskell.

One thing to keep in mind that may be confusing for beginners coming from the imperative world: We are not applying the new `runDeckM` computation to anything, we are defining a *new function* on the fly using *lambda notation* (the `\deck ->` part, where symbol "\" stands for Greek lambda). We can rewrite it without lambda in the following way to make it a bit easier to grasp:

```
fmap f (DeckM {runDeckM = runDeckOld}) = DeckM { runDeckM = runDeckNew }
   where
   runDeckNew deck = let (x , newDeck) = runDeckOld deck
                     in (f x, newDeck)
```

Thus, our `runDeckNew` function takes a `deck::Deck` and returns whatever comes after `in` on the right, so a pair `(f x, d)` – that *must* have the type `(b, Deck)` for the whole thing to typecheck. Does it? Well, what we do before `in` is apply the `runDeckOld` computation to the `deck::Deck`, and since `runDeckOld` is just `runDeckM` from the `DeckM a` value on the left, it means that it produces `(a, Deck)`. This in turn means that `d::Deck` and `x::a`. Then, `(f x) :: b` since `f :: a -> b` and everything typechecks!

The only thing that's left is to check that the **Functor** laws are being preserved – which I leave to you. It's much easier than writing the actual definition above.

CHAPTER 6 O, MONAD, HELP ME COMPOSE!

Let's move on to **Applicative**:

```
instance Applicative DeckM where
    pure x = DeckM { runDeckM = \d -> (x, d) }
    (<*>) (DeckM { runDeckM = rf} ) (DeckM {runDeckM = rx}) =
        DeckM {
            runDeckM = \d -> let (f, d')  = rf d
                                 (x, d'') = rx d'
                             in  (f x, d'')
        }
```

To make something an instance of **Applicative,** we need to define pure, which lifts a pure value into our type function (in this case DeckM), quite trivial in this case: we simply return a function that takes a deck and returns this deck together with our value – and one of <*> or liftA2. Since we defined liftA2 previously, let's practice defining <*>, which should have the following type signature: DeckM (a -> b) -> DeckM a -> DeckM b. This means that we have a function a->b "lifted" into DeckM and we have to somehow apply it to the value a lifted into DeckM to get DeckM b as a result. If you followed the explanation for fmap above, unpacking the <*> definition shouldn't be difficult for you as it follows a similar logic. We define a new computation runDeckM, which takes a deck::Deck, apply runDeckM from the first argument to extract the function f::a->b from inside DeckM, then apply runDeckM from the second argument to the new deck (d') to thread the state and extract x::a, and then return a pair (f x, d").

By now, defining a **Monad** instance for **DeckM** should be easy for you: we need to write operator (>>=) :: DeckM a -> (a -> DeckM b) -> DeckM b, which will allow us to chain monadic **DeckM** computations together. It is in fact easier than **Applicative**:

```
instance Monad DeckM where
    (>>=) :: DeckM a -> (a -> DeckM b) -> DeckM b
```

125

CHAPTER 6   O, MONAD, HELP ME COMPOSE!

```
(>>=) (DeckM {runDeckM = rd}) f = DeckM { runDeckM = \d ->
  let (x, d') = rd d in runDeckM (f x) d' }
```

The only notable difference here is that f has the type a -> DeckM b, so the part after in first applies f to the extracted argument to get DeckM b, and then we apply runDeckM from this new DeckM b to the deck (d') that's being threaded through our computations. Dwell on it until you fully understand everything that's happening and don't forget to check that the laws hold!

Phew! If you were able to follow the above train of thought, you will have no problems with the rest of Haskell. If you found it difficult, it's not a problem at all, since as we mentioned above, for practical Haskell programming you do not really need to define new **Monad** instances (*unless you are writing new complex libraries for others to use*) – in most of the cases using the existing libraries is more than enough! So don't despair, but read the next sections of this chapter to learn how to use the existing Monads without digging into their complicated internals.

So why go through all this trouble? What does making **DeckM** a **Monad** give us? As we will soon see, we "invented" a so-called **State Monad**, which allows you to thread any kind of "global state" through your computations without resorting to mutable variables. In our case, the state is simply the dealer's deck, but we can make it anything we want. Before we do, let's finish the **DeckM** example and provide two helpful functions that manipulate our state:

```
getDealerDeck :: DeckM Deck
getDealerDeck = DeckM {runDeckM = \d -> (d,d)}

putDealerDeck :: Deck -> DeckM ()
putDealerDeck deck = DeckM {runDeckM = \d -> ((),deck)}
```

The first one returns the state (dealer's deck) as a result of our computation, and the second puts the new state into our thread. Now we can rewrite the dealCards and countCards functions from above as follows:

```
dealCardsM :: Int -> DeckM Deck
dealCardsM n = getDealerDeck >>= \deck ->
                  let newDeck = drop n deck
                      plDeck  = take n deck
                  in putDealerDeck newDeck >> pure plDeck

countCardsM :: DeckM Int
countCardsM = getDealerDeck >>= pure . length
```

The second is straightforward, while the first manipulates the state, so we have to get it first by running getDealerDeck, then dealing cards, and then putting it back before returning the dealt cards as pure plDeck. Now you can combine these (and any other computations of the type a -> DeckM b!) monadically like this:

ghci> runDeckM (putDealerDeck fullDeck >> dealCardsM 3) []
([2♣,3♣,4♣],[5♣,6♣,7♣,8♣,9♣,10♣,J♣,Q♣,K♣,A♣,2♠,3♠,4♠,5♠,
6♠,7♠,8♠,9♠,10♠,J♠,Q♠,K♠,A♠,2♦,3♦,4♦,5♦,6♦,7♦,8♦,9♦,10♦,
J♦,Q♦,K♦,A♦,2♥,3♥,4♥,5♥,6♥,7♥,8♥,9♥,10♥,J♥,Q♥,K♥,A♥])
ghci> runDeckM (putDealerDeck fullDeck >> dealCardsM 3 >> countCardsM) []
(49,[5♣,6♣,7♣,8♣,9♣,10♣,J♣,Q♣,K♣,A♣,2♠,3♠,4♠,5♠,6♠,7♠,8♠,
9♠,10♠,J♠,Q♠,K♠,A♠,2♦,3♦,4♦,5♦,6♦,7♦,8♦,9♦,10♦,J♦,Q♦,K♦,
A♦,2♥,3♥,4♥,5♥,6♥,7♥,8♥,9♥,10♥,J♥,Q♥,K♥,A♥])

In the first case, we put the fullDeck into our state and then deal three cards from it; in the second we then count the remaining dealer's deck cards. Isn't it beautiful? We don't pass the state around explicitly anymore, but can still manipulate it! From the above, the point of naming our computation runDeckM also becomes clear – it executes whatever chained monadic action we give it along with the initial state! It also means that we can simplify the above calls to get rid of the initial putDealerDeck calls by simply providing fullDeck as the starting state: runDeckM (dealCardsM 3 >> countCardsM) fullDeck.

We will see the full power of the **State Monad** in the final Blackjack game code, but this should give you an initial idea – along with the sense of accomplishing a non-trivial task of defining a new Monad from scratch!

Monad Typeclass and Basic Monads

Let's look at the full definition of the **Monad** typeclass, which becomes very easy thanks to our in-depth treatment of **Applicative** in the last chapter:

```
class Applicative m => Monad (m :: Type -> Type) where
    (>>=) :: m a -> (a -> m b) -> m b
    (>>) :: m a -> m b -> m b
    (>>) = (*>)
    return :: a -> m a
    return = pure
```

As you can see, we only need to define operator >>= to introduce a new Monad; the remaining functions are fully defined in terms of Applicative. While >> makes sense aesthetically when you write monadic code (if you recall, it simply chains computations together, produces the effects, and discards the results), return is there purely for historic reasons as Applicative was added to Haskell long after Monad. I personally prefer pure wherever possible.

Monad typeclass must satisfy certain laws, just as Functor and Applicative. These are:

CHAPTER 6   O, MONAD, HELP ME COMPOSE!

- Left identity: pure a >>= comp = comp a. This is quite trivial if you think about the corresponding type signatures.

- Right identity: comp >>= pure = comp. Again, think about the type signatures and it will become clear.

- Associativity: comp1 >>= (\x -> comp2 x >>= comp3) = (comp1 >>= comp2) >>= comp3. There is another lambda function definition, but in essence it just says that monadic computations have to be associative (if you read Chapter 3, you may think if there's a separate category somewhere – indeed there is, and we will discuss it in Chapter 13).

These laws imply the following relations between monadic and Functor/Applicative operations:

- pure = return

- m1 <*> m2 = m1 >>= (\x1 -> m2 >>= (\x2 -> pure (x1 x2)))

- fmap f xs = xs >>= pure . f

Can you show it is true?

Also, if you went through the pain of defining Functor–Applicative–Monad for the DeckM type from the previous section, you may think "WHY?! We could have simply defined Monad and then get Functor and Applicative automatically?! And instead you made us go through this pain?!" Yes. But that's the only way to learn, isn't it? You can take shortcuts later.

Monads are ubiquitous in Haskell and serve as the main way to structure real-world programs by chaining or composing monadic computations together – see Figure 6-3.

129

CHAPTER 6 O, MONAD, HELP ME COMPOSE!

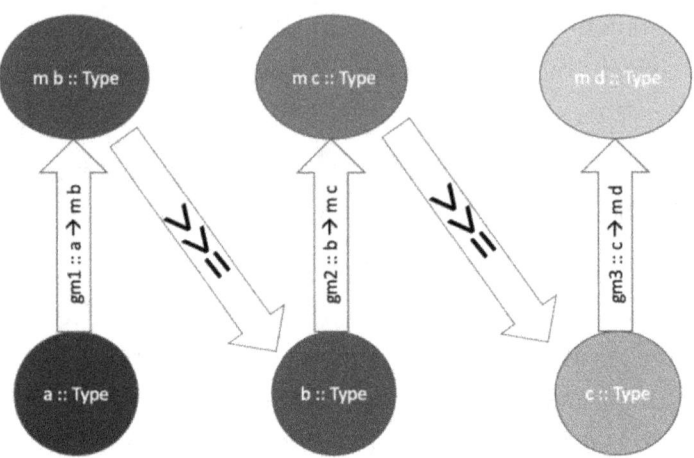

Figure 6-3. *Monadic computations composed together with the help of >>= operator, but with values always remaining "captured" inside the monad*

We will gradually explore many of them, but let's start with the easiest Monad that we already know – **Maybe**.

EXERCISE

Define Applicative and Monad instances for Maybe (easy) and List (harder). How many ways can you think of to make List a monad?

Maybe type is arguably the simplest Monad you can imagine. I am sure you had no issues writing an **Applicative** instance for it, and to define a **Monad** we need only to define the (>>=) operator:

```
(>>=) :: Maybe a -> (a -> Maybe b) -> Maybe b
Nothing  >>= f = Nothing
(Just x) >>= f = f x
```

If a computation receives Nothing as an argument, it returns Nothing. If it receives Just x, it takes x as an argument, does what it's supposed to, and returns Maybe b. Trivial. If you look carefully at this definition, you will realize that if you chain monadic computations for Maybe, it will only execute in full if all the computations return Just somethings. Even with one Nothing – and that's it – the overall result is Nothing. Remember our safeDivide function from Chapter 1?

```
safeDivide :: Double -> Double -> Maybe Double
safeDivide _ 0 = Nothing
safeDivide x y = Just (x/y)
```

Now we can *chain* these computations together:

```
ghci> safeDivide 4 5 >>= safeDivide 3 >>= safeDivide 28
Just 7.466666666666667
ghci> safeDivide 4 5 >>= safeDivide 0 >>= safeDivide 28
Nothing
```

This is of course a trivial example, but you get the idea of how your code becomes much more concise and easier to read thanks to the >>= operator. Just like performing magical spells in order!

As we by now repeated several times, Monad is just another typeclass and as such an abstraction that adds certain *structure* to our types while being a logical development of Functor and Applicative. And just like a circle is an abstraction both for the sun and for the vinyl record – two very different things – Monad is an abstraction for many quite different types and computation styles, which gives us a very nice way to *compose* these computations while providing very nice isolation options. Let's look at the most common monads that you will be using in almost any real-world Haskell application.

CHAPTER 6   O, MONAD, HELP ME COMPOSE!

Reader–Writer–State Triple

Reader, Writer, and State monads are the most common abstractions you will be using in your programs – along with IO, of course, without which no input–output is possible in Haskell. Here the situation is slightly complicated, because in Haskell there are two libraries that provide RWS monads: *transformers* and *mtl*. The latter is built on top of the former and provides additional typeclasses that help you structure your code, but uses so called *functional dependencies* in their definitions, which is a slightly advanced topic. But since we have defined Functor-Applicative-Monad for the DeckM type above, this is nothing to be afraid of. The monad that we have defined for DeckM is a type of a State monad, since it allows us to compose computations while threading "global" state through all of them. In our simplified card playing example, the state was simply one Deck, but even in the final Blackjack program, it will be much more complex. In a real-world app with a database, web API, and complex business logic layer, even more so. So how do we use any kind of state in our application?

The State Monad

Let's look at the interface for the State monad itself – as this will likely be the most common monad you'll use in your code, along with IO:

```
type State s = StateT s Identity
```

Don't concern yourself with the right side of this equation yet, we will discuss it in detail in the Chapter 7 where we introduce Monad transformers. For practical purposes, all we need to do to create a State monad with our own type of state s is to write State s. Let us rewrite the DeckM example from above using this interface along the way to make things practical. Since DeckM used Deck for the state, we can redefine it as follows:

```
type DeckM' = State Deck
```

This will automatically create a State monad with type **Deck** used to keep the state. Now we can also rewrite the functions dealCardsM and countCardsM:

```
dealCardsM' :: Int -> DeckM' Deck
dealCardsM' n = get >>= \deck ->
                  let newDeck = drop n deck
                      plDeck  = take n deck
                  in put newDeck >> pure plDeck

countCardsM' :: DeckM' Int
countCardsM' = get >>= pure . length
```

Contrast these definitions with the ones we came up with above – do you notice the difference? We are using get and put polymorphic functions that are provided to us by the mtl package now to access our state instead of the ones we had to write by hand. Then, we can use other functions from the mtl **State** interface to execute our computations:

```
ghci> runState (dealCardsM' 3 >> countCardsM') fullDeck
(49,[5♣,6♣,7♣,8♣,9♣,10♣,J♣,Q♣,K♣,A♣,2♠,3♠,4♠,5♠,6♠,7♠,8♠,
9♠,10♠,J♠,Q♠,K♠,A♠,2♦,3♦,4♦,5♦,6♦,7♦,8♦,9♦,10♦,J♦,Q♦,K♦,
A♦,2♥,3♥,4♥,5♥,6♥,7♥,8♥,9♥,10♥,J♥,Q♥,K♥,A♥])
ghci> execState (dealCardsM' 3 >> countCardsM') fullDeck
[5♣,6♣,7♣,8♣,9♣,10♣,J♣,Q♣,K♣,A♣,2♠,3♠,4♠,5♠,6♠,7♠,8♠,9♠,
10♠,J♠,Q♠,K♠,A♠,2♦,3♦,4♦,5♦,6♦,7♦,8♦,9♦,10♦,J♦,Q♦,K♦,A♦,
2♥,3♥,4♥,5♥,6♥,7♥,8♥,9♥,10♥,J♥,Q♥,K♥,A♥]
ghci> evalState (dealCardsM' 3 >> countCardsM') fullDeck
49
```

As you can see, besides runState, which is a replacement for our previously defined runDeckM, there are two more convenience functions execState and evalState, where the former executes all computations

CHAPTER 6 O, MONAD, HELP ME COMPOSE!

and returns the final state, while the latter executes all computations and returns the final value. Here are the most important functions from the State Monad interface, given a State monad m s:

```
-- State manipulations:

-- get the state from the state monad:
get :: m s
-- put the new state into the monad:
put :: s -> m ()
-- modify state with the function s -> s:
modify  :: (s -> s) -> m ()
-- strict version of modify:
modify' :: (s -> s) -> m ()
-- get some components of the state (e.g., specific field of a
record) using projection function (s->a):
gets :: (s -> a) -> m a

-- Executing monadic computations:
runState  :: State s a -> s -> (a, s)
evalState :: State s a -> s ->  a
execState :: State s a -> s ->  s
```

Study them carefully as you will be using them daily in your Haskell programs. You have seen the usage examples for most of them above, with the exception of modify and gets, but their purpose should be self-explanatory from the type signature and the comments. Using modify, we can rewrite the dealCardsM' function even more concisely:

```
dealCardsM' n = get >>= \deck ->
                let plDeck = take n deck
                in modify (drop n) >> pure plDeck
```

```
-- or, noticing that we can get rid of plDeck:
dealCardsM' n = get >>= \deck -> modify (drop n) >> pure $ 
take n deck

-- or, rewriting in the logical order of operations:
dealCardsM' n = get >>= (pure . take n) >>= \plDeck -> (modify 
(drop n) >> pure plDeck)
```

Isn't this beautiful and just like our chaining spells in Chapter 1? Every computation is some logical action or transformation, and they are being glued together by the monadic operators >>= and >>.

As for gets, it will come in very handy when your state is a record with lots of fields and you need to access just one or two of them.

> **EXERCISE WITH STATE**
>
> Come up with an example of some stateful computation that you would like to perform. Make a State monad out of it via State s type function and solve your problem.

The Reader Monad

If the State monad is about threading state through a sequence of computations, the Reader monad is all about passing around "shared context" without explicitly passing it in each function. Just like in State, where the state flows through computations secretly, Reader will let us access some immutable environment behind the scenes.

This is extremely helpful in scenarios where you need to use some configuration settings, logging context, or other global but immutable data across your application without cluttering your function signatures with extra parameters. Imagine writing functions where you always had to explicitly pass some configuration in and out – that would become cumbersome quickly. The Reader monad allows us to abstract this away.

Let's start by taking a look at the basic definition of the Reader monad:

```haskell
type Reader r a = ReaderT r Identity a
-- or with so called "eta-reduction":
type Reader r = ReaderT r Identity
```

Just like with State, you don't need to worry too much about the right-hand side yet. The key part to focus on is that Reader r gives you a monad where r is the type of the environment you're working with and a is your return type. In other words, Reader r a represents a computation that can access some environment of type r and ultimately produce a result of type a.

Even though in our very simplified "Blackjack" example we don't really need the Reader monad, it is easy enough to imagine a scenario where we would – e.g., if we wanted to make our game multiplayer and we needed to share some global settings for the server, connections, etc. It would be very cumbersome to pass this configuration everywhere explicitly, so making them "hidden" inside a Reader monad is a great solution.

Below we define a type synonym ConfigM to represent computations that operate in the context of a configuration of type Config:

```haskell
-- Our shared environment, a configuration record
data Config = Config
  { baseUrl        :: String
  , maxConnections :: Int
  } deriving (Show)

-- A Reader monad that provides access to Config
type ConfigM = Reader Config
```

CHAPTER 6 O, MONAD, HELP ME COMPOSE!

Now, let's write a function that reads from this shared Config without explicitly passing it as a parameter to each function.

```
-- Get the base URL from the configuration
getBaseUrl :: ConfigM String
getBaseUrl = asks baseUrl
```

Whoa, what's this asks thing? It's a helper function that comes with the Reader monad! It allows you to extract a piece of the environment by applying a projection function (Config -> a) to it. Here, we're using it to grab the baseUrl field from the Config record.

Let's create another function that also reads from the config:

```
-- Get the maximum number of connections from the configuration
getMaxConnections :: ConfigM Int
getMaxConnections = asks maxConnections
```

Now that we have some computations inside our Reader monad, how do we actually run them? Just like with State, there's a function to execute the Reader monad and pass in the environment:

```
runReader :: Reader r a -> r -> a
```

You provide the environment r and it executes the computation, returning the result. Let's see this in action:

```
-- Sample configuration
myConfig :: Config
myConfig = Config "https://api.example.com" 10

-- Running Reader computations
ghci> runReader getBaseUrl myConfig
"https://api.example.com"

ghci> runReader getMaxConnections myConfig
10
```

Notice how neat and clean this is! We didn't have to pass the config explicitly into getBaseUrl or getMaxConnections. Instead, we just supplied it once through runReader, and both functions seamlessly accessed it from the shared environment.

Let's take things up a notch. Suppose we need a function that executes several operations within the same Reader monad context, combining results from multiple computations:

```
-- Combine the base URL and max connections into a
formatted string
formatConfigInfo :: ConfigM String
formatConfigInfo = do
  url  <- getBaseUrl
  maxC <- getMaxConnections
  pure $ "Base URL: " ++ url ++ ", Max Connections: " ++
  show maxC
```

In the code above, we are using so-called do notation, which we will formally discuss in the next chapter. In essence, in the first line, the "<-" operator binds the result of the getBaseUrl monadic computation to the "url" "variable," likewise for the "maxC," and then we can use them both in the third line. We can now run this combined computation:

```
ghci> runReader formatConfigInfo myConfig
"Base URL: https://api.example.com, Max Connections: 10"
```

See how effortlessly url and maxC are accessed from the environment without us explicitly threading Config through the calls? That's the beauty of the Reader monad.

Here are some key functions that you'll be using frequently when working with Reader monads:

```
-- Retrieves the entire environment:
ask :: m r
```

CHAPTER 6 O, MONAD, HELP ME COMPOSE!

```
-- For example:
showConfig :: ConfigM Config
showConfig = ask
-- This gives you the whole environment inside the monad.
-- A more specific version of ask that applies a projection
function:
asks :: (r -> a) -> m a
```

We've already seen this in our previous examples, where we used it to extract specific fields from Config.

```
-- Temporarily modify the environment for a given computation:
local :: (r -> r) -> Reader r a -> Reader r a
```

This function allows you to "override" the environment for a specific computation. For example, you could temporarily change the baseUrl for a particular call:

```
customBaseUrl :: ConfigM String
customBaseUrl = local (\config -> config { baseUrl = "https://custom-url.com" }) getBaseUrl
```

Now, running this code:

```
ghci> runReader customBaseUrl myConfig
"https://custom-url.com"
```

The baseUrl is changed just for that one computation, while the original config remains intact elsewhere.

EXERCISE WITH READER

Come up with three very common scenarios where the Reader monad would be useful.

CHAPTER 6   O, MONAD, HELP ME COMPOSE!

That's the Reader monad for you! It's an elegant tool to manage shared, immutable context across computations without the headache of passing parameters everywhere. Just like casting a spell, you can access the environment silently without revealing the magic behind the scenes.

The Writer Monad

We've seen how the State monad helps us thread mutable state through computations and how the Reader monad lets us access shared read-only context. But what if we wanted to log information as we go or accumulate auxiliary data alongside our main computations, without explicitly passing that logging structure around? This is where the Writer monad comes in.

The Writer monad allows us to "secretly" carry an additional value alongside our results – typically some kind of log or output that we can accumulate throughout our computation. This is incredibly helpful for scenarios like collecting debug information, building up a report, or gathering statistics while performing some other task.

The core idea of the Writer monad is that you have a computation that produces both a result and a "log" of some sort. The log values are accumulated (or "appended") as the computation proceeds. This is made possible by the fact that the log is associated with a **Monoid**, which if you recall from the previous chapters gives it an associative binary operation (mappend) and a neutral element (mempty).

Here's the type definition of the Writer monad:

```
type Writer w = WriterT w Identity
```

Again, don't worry about the WriterT part quite yet. The important thing for now is that Writer w a represents a computation that produces two things:

- A result of type a
- An accumulated "log" of type w, where w is expected to be a Monoid

For example, `Writer [String] Int` would represent a computation that returns an Int along with a list of strings as logs.

Before diving into examples, here are some key functions you'll be using with the Writer monad:

```
-- Appends a given "log" to the accumulated output:
tell :: Monoid w => w -> Writer w ()

-- Creates a Writer value from a result and a log:
writer :: (a, w) -> Writer w a

-- Runs a Writer computation and returns both the result and
the accumulated log:
runWriter :: Writer w a -> (a, w)

-- Similar to runWriter, but returns only the accumulated log,
discarding the result of the computation:
execWriter :: Writer w a -> w

-- Like runWriter, but returns only the result, discarding
the log:
evalWriter :: Writer w a -> a
```

Let's revisit the card-related types we explored in the "The State Monad" section and imagine that we want to perform some computations while logging the actions we're taking. For instance, we might want to deal cards while logging the actions ("Dealt 3 cards" and so on) taken during execution.

CHAPTER 6   O, MONAD, HELP ME COMPOSE!

First, we'll define a type alias for the Writer monad where the log is a list of strings, which will represent each action we perform:

```
-- Writer monad for logging actions (each action will be
a String)
type DeckMLog = Writer [String]
```

Now, let's redefine our previous dealCardsM function from the "The State Monad" section, but this time we'll use the Writer monad to log the actions we perform:

```
dealCardsW :: Int -> Deck -> DeckMLog [Card]
dealCardsW n deck =
  let newDeck = drop n deck
      plDeck  = take n deck
  in writer (plDeck, ["Dealt " ++ show n ++ " cards."])
```

In the above code

- We use the writer function to return a tuple: the result of the computation (plDeck) and the log (a list of strings).
- The log consists of a message about how many cards were dealt.

Let's write another function, countCardsW, that logs the number of cards remaining in the deck:

```
countCardsW :: Deck -> DeckMLog Int
countCardsW deck = writer (length deck, ["Counted cards: " ++ show (length deck)])
```

Now, let's combine these two computations using the Writer monad:

```
playRoundW :: Int -> Deck -> DeckMLog (Deck, Int)
playRoundW n deck = do
```

```
  dealtCards <- dealCardsW n deck
  remainingCards <- countCardsW (drop n deck)
  pure (drop n deck, remainingCards)
```

Just like with State and Reader, we need a way to execute our Writer computations. As mentioned earlier, we can use runWriter to get both the result and the accumulated log:

```
ghci> runWriter (playRoundW 3 fullDeck)
((remainingDeck, 49), ["Dealt 3 cards.", "Counted cards: 49"])
```

But often, we're interested in either the result or the log. So we can use evalWriter to get just the result:

```
ghci> evalWriter (playRoundW 3 fullDeck)
(remainingDeck, 49)
```

Or we can use execWriter to get just the log:

```
ghci> execWriter (playRoundW 3 fullDeck)
["Dealt 3 cards.", "Counted cards: 49"]
```

As you can imagine, the power of the Writer monad lies in its ability to log information as you go, combining logs from multiple computations seamlessly. For example, let's add some more logging to the playRoundW function:

```
playRoundW :: Int -> Deck -> DeckMLog (Deck, Int)
playRoundW n deck = do
  tell ["Starting a new round..."]
  dealtCards <- dealCardsW n deck
  tell ["Calculating remaining cards..."]
  remainingCards <- countCardsW (drop n deck)
  tell ["Round completed!"]
  pure (drop n deck, remainingCards)
```

Now the log will include additional messages about the progress of the round:

```
ghci> execWriter (playRoundW 3 fullDeck)
["Starting a new round...",
 "Dealt 3 cards.",
 "Calculating remaining cards...",
 "Counted cards: 49",
 "Round completed!"]
```

This is a powerful pattern when you need to track what's happening in your computation, especially for debugging or generating reports.

The Writer monad isn't restricted to logging with strings. Since Writer works with any type that is a Monoid, you can use it to accumulate all kinds of data. For example, you could accumulate a list of results, a sum of numbers, or combine any custom data structure that forms a Monoid.

For instance, if we wanted to track the number of cards dealt as a sum, we could use the Sum Monoid:

```
import Data.Monoid (Sum(..))

type DeckMSum = Writer (Sum Int)

dealCardsSum :: Int -> Deck -> DeckMSum [Card]
dealCardsSum n deck = writer (take n deck, Sum n)
```

Now, running this computation will give us both the result and the total number of cards dealt:

```
ghci> runWriter (dealCardsSum 3 fullDeck)
([dealtCards], Sum 3)
```

- The Writer monad allows us to accumulate auxiliary data (like logs) alongside our main computation, without explicitly passing that data around.

- The output of a Writer computation is a Monoid, which means it can be combined using mappend and has an identity element (mempty).

- Common use cases include logging, debugging, building reports, or accumulating any additional data alongside the result of a computation.

EXERCISE WITH WRITER

Imagine you are developing a simplified banking system where each transaction (deposit or withdrawal) needs to be logged. Define a Transaction type to represent actions taken, and use the Writer monad to keep a log of all transactions as they are processed. Accumulate the total balance and log each transaction along the way.

Conclusion

The Writer monad is another powerful tool in Haskell's monadic toolbox. It allows you to write clean, modular code that logs or accumulates data invisibly and automatically. Whether you're building a logging system or combining other auxiliary data like sums or statistics, Writer helps you do so with ease and elegance.

The question may arise, but why do I need Reader and Writer when I can achieve pretty much the same result with State? You may, but separation of abstractions is a very good practice for your code maintainability. Having one huge State structure in your program is cumbersome, plus you never know if you accidentally modified some values that are supposed to be read-only and so on. Hence, combining Reader, Writer, and State lets you write good, modular, maintainable code.

CHAPTER 6   O, MONAD, HELP ME COMPOSE!

You may also wonder, but how do I combine several different monads in one program? That is what Monad transformer stacks are for, and we are very close to finally discussing them and starting to write "real-world" programs. But before we do, let's dive into the IO, which is also incidentally a monad.

CHAPTER 7

Input, Transformer Stack, Output

What a road it has been so far! We have laid a solid foundation of types, functions, and various useful abstractions built in a neat hierarchy of typeclasses such as Functor–Applicative–Monad. Now, finally, we are very close to starting to build real-world programs. The two remaining puzzle pieces are the so-called IO Monad, which will allow us to perform input/output operations, and Monad transformer stacks – the beautiful way to structure Haskell programs of arbitrary complexity. They are the topic of this chapter.

Let us start with the IO Monad. Input/output is a difficult problem from the mathematical point of view. We have been carefully building our *pure* **Hask** world – mathematical world without side effects, where each function behaves predictably and we can guarantee mathematically that our programs do what we intend them to do. As soon as we introduce input/output operations, we introduce uncertainty and utter unpredictability, associated with our messy human realm. Thus, we need some way to isolate our pure, well-behaved functions and types from the messy, random, unpredictable IO operations. Designers of the Haskell

CHAPTER 7   INPUT, TRANSFORMER STACK, OUTPUT

language found that Monad abstraction works very well for this purpose – if you recall from the previous chapter, once a certain computation is "wrapped" inside a Monad, it can never escape.[1]

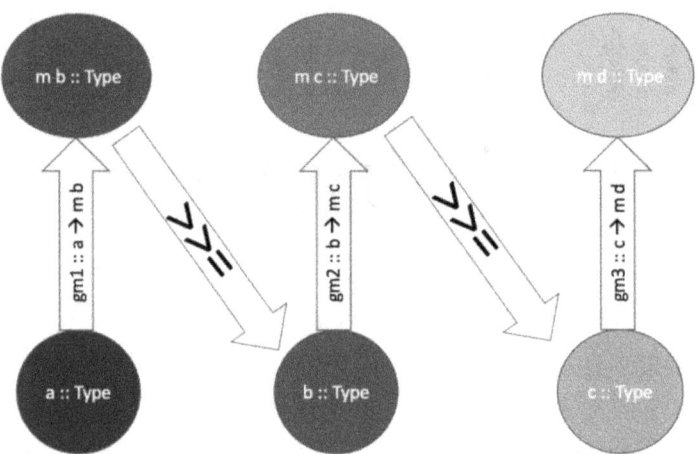

Figure 7-1. *Monadic computations composed together with the help of >>= operator, but with values always remaining "captured" inside the monad*

We can chain as many monadic computations together as we want with the help of >>= operator, but the result of such a chain will always be wrapped inside the monad. Why is it good for us to handle IO? Because we can design most of our types and functions for our problem domain as pure, getting all the mathematical benefits (practically formulated concisely as "if it compiles it runs"), and then we can simply "lift" our purely defined world into the messy IO monad with all the side effects

---

[1] Unless we explicitly provide an "escape hatch" as designers of a certain monadic type. For example, there is such an "escape hatch" for Maybe monad – by simply pattern-matching any Maybe value like "Just x," we can get access to the actual "x" value, wrapped inside the Maybe monad. But as we will see it is not mandatory, and we can just as easily *not* provide it, and then our values will forever live inside the monad – exactly what we need with IO.

CHAPTER 7   INPUT, TRANSFORMER STACK, OUTPUT

and uncertainty, but we can never make the reverse transformation – no value that lives in the IO monad can make it back in the pure world to contaminate it!

This may seem a bit confusing at first, so let me give you a simple example from the imperative language world. There, we can write something like using, for instance, terrible, terrible JavaScript:

```
function impureSquare(x) {
    const y = x * x;
    console.log("Result is", y);
    return y;
}
```

Here, we have an ugly spaghetti of confusion that is oh so prevalent in the imperative languages. We have a pure mathematical function, squaring a number, but we put side effects of writing something to the console inside it! And while "writing to the console" is a relatively harmless side effect, this could be altering the memory, which leads to much, much worse consequences. This can never happen in Haskell. You cannot put *any* IO operation inside any pure function – and this gives you this beautiful separation of concerns and maintainability that is so difficult to achieve in other languages, but you *can* put (or "lift") any pure function into the IO monad! The trend of the recent years is that even imperative languages are trying to implement best practices from Haskell, and even JavaScript with popular libraries such as React is trying to promote designing most of your code with immutable, pure functions, but of course since the language still allows you to write any kind of messy spaghetti of mixing effects with pure functions with events and what have you, it is very rarely followed. Haskell imposes certain restrictions on you, which may seem a bit cumbersome at first if you are coming from the imperative world, but they make you a much better programmer in the long run.

CHAPTER 7   INPUT, TRANSFORMER STACK, OUTPUT

But let's get back to the IO monad. Recall the Monad typeclass definition (omitting secondary functions):

```
class Applicative m => Monad (m :: Type -> Type) where
    (>>=) :: m a -> (a -> m b) -> m b
    (>>) :: m a -> m b -> m b
    pure :: a -> m a
```

As mentioned above, >>= allows us to chain monadic computations together, while pure (borrowed from Applicative and called "return" in Monad, but we prefer to write "pure" where possible) "lifts" a value inside our monad. In essence, this is all we need to know to start doing IO – since it is a monad, it supports all the operations above – but since we are curious individuals, let's take a peek under the hood.

IO Monad is defined as follows in Haskell:

```
newtype IO a = IO (State# RealWorld -> (# State# RealWorld, a #))
```

Ouch. Looks confusing. Let's ignore the "#" signs for now – we'll have a discussion of what they mean down the road, but they are not essential for our conceptual discussion – and rewrite it as

```
newtype IO a = IO (State RealWorld -> (State RealWorld, a))
```

If it looks familiar, it should – remember how we made a DeckM type an instance of Monad typeclass in the last chapter and we realized that it is in fact a State Monad? DeckM was defined as follows:

```
newtype DeckM a = DeckM { runDeckM :: Deck -> (a, Deck)}
-- or, if we unwind the named record, turn it into a tuple and
switch the places on the right:
newtype DeckM a = DeckM ( Deck -> (Deck, a) )
```

Don't you just love finding the same abstractions among seemingly different entities? The above similarity tells us, since DeckM is a State monad and Deck is the state that we pass from computation to computation, that IO is also a State monad, where the state is "State RealWorld"! So, in essence, it literally tells us that in every IO computation *we are changing the state of the real world*! And that's exactly what happens with IO – either the user types something in, or we display something on the screen, or we change the state of memory by updating mutable variables, and so on – whatever we do, we are affecting the real world. Making IO into a type of State monad where the state is the state of the whole world is so beautiful, don't you think?

However, there are of course notable differences between IO monad and the State Monad typeclass we discussed in the previous chapter. The first one is that unlike in any State Monad, we can extract neither a value nor the state from inside the IO monad. Once you lifted a value or a computation into IO, there is no way back. This is done exactly because we want to isolate our pure world from all the IO side effects as described above. The second is that "State RealWorld" is not a "real" state, but rather some internal Haskell compiler "magic" that has to do with how IO monad is implemented under the hood – and we won't go into those very technical details here. All of this discussion is to satisfy our curiosity a bit and absolutely unnecessary to *use* the IO monad. So let's get practical.

Remember the example from the very first chapter where we solved a problem of calculating an average of the list of numbers?

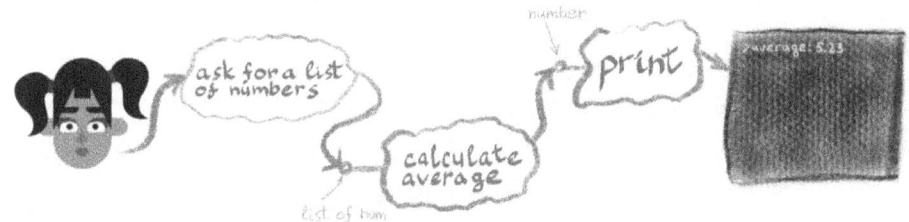

Figure 7-2. *Functional way to solve problems*

CHAPTER 7 INPUT, TRANSFORMER STACK, OUTPUT

The resulting translation of Figure 7-2 into actual Haskell code was the following:

```
main = ask >> getLine >>= toList >>= calcAverage >>=
printResult
      where calcAverage l = pure $ (fromIntegral $ sum l) /
      (lengthF l)
            printResult n = putStrLn $ "Average of your list
            is: " ++ (show n)
            ask = putStrLn "Enter list of numbers, e.g.
            [1,3,4,5]:"
            lengthF = fromIntegral . length
            toList :: String -> IO [Int] = pure . read
```

Good news is now we can finally completely decode it! The first line, it turns out, is merely chain monadic computations in the IO monad, just like in Figure 7-1, which are defined below in the "where" block:

- ask: Simply puts a prompt to the terminal.

- getLine: Reads an input line from the terminal.

- toList: Tries to convert whatever was entered to a list of numbers. This one is a bit tricky and is a composition of two functions:

  - read: Tries to convert a string into an [Int] value

  - pure: "Lifts" this value into the IO monad

- calcAverage: Calculates the average of the list of numbers that we received.

- printResult: Shows the result with some comments.

This also hints at how to write actual Haskell programs – you have to define a function called "main," which always has type IO (in other words, is an IO computation), and it will serve as an entry point to your program. It will always be a chain of other IO computations. In practice, as we discussed repeatedly, first, you have to design your pure types and functions, then think how to structure your types and functions into various monads such as Reader–Writer–State, and then somehow combine all of it with the ability to perform IO actions using IO monad. The way to do it is by using Monad transformer stacks, which we will discuss very soon. But for now, let's finish with the IO monad overview.

IO Monad allows you to do all kinds of interactions with the external world – writing to the terminal, reading from the terminal, working with the file system, working with the operating system, adjusting memory, working with mutable variables, accessing networking operations – everything you may expect. We will not be covering all the available functions here, as you can easily find all of them on hackage.haskell.org – the main repository of Haskell libraries. As we are building different programs in the second part of this book, we will cover some of those functions, but as the main goal of this book is to teach you to easily read type signatures and understand what functions do based on that, I am sure by now you are fluent enough in working with Hackage in this way.

Do Notation

Another topic related to the IO monad is so-called do notation. In fact, you can us it with any monad, but it is most commonly associated with IO. Personally, I prefer to write any monadic code using >>= and >> operators as much as possible. In my opinion, it captures the spirit of what a monad is much better and looks more beautiful. However, in some cases, especially when we need to stitch the result of many computations into a

CHAPTER 7   INPUT, TRANSFORMER STACK, OUTPUT

sophisticated multiparameter computation, it might be easier to use "do notation" – which is only syntactic sugar for the aforementioned monadic operators. We have already seen some examples of the "do notation" usage in the previous chapter. Now let's decode it in detail.

Let's quickly review how we typically work with monads using the **bind** operator (>>=). The bind operator is the key mechanism that lets us sequence monadic computations. Here's a quick refresher on its type signature:

```
(>>=) :: Monad m => m a -> (a -> m b) -> m b
```

- It takes a monadic value m a (a value wrapped in a monad) and a function (a -> m b) (a function that takes a normal value and returns a monadic value).

- It applies the function to the unwrapped value and chains the two computations together.

Let's look at Maybe to see this in action. Imagine we have a chain of computations where each step might fail (return Nothing), using the example from the previous chapters:

```
safeDivide :: Double -> Double -> Maybe Double
safeDivide _ 0 = Nothing
safeDivide x y = Just (x / y)

exampleWithoutDo :: Maybe Double
exampleWithoutDo =
  safeDivide 10 2 >>= \result1 ->
  safeDivide result1 5 >>= \result2 ->
  safeDivide result2 4
```

Here, we're chaining three safe divisions, and each step may fail. The >>= operator lets us pass the result of each successful computation to the next.

CHAPTER 7   INPUT, TRANSFORMER STACK, OUTPUT

Now, that's fine for small computations, but as the chains get longer, using >>= becomes awkward quickly. This is where **do notation** swoops in like a hero – or a villain, depending on your taste. Here's how the above code would look using do notation:

```
exampleWithDo :: Maybe Double
exampleWithDo = do
  result1 <- safeDivide 10 2        -- Perform the first safe
                                       division
  result2 <- safeDivide result1 5   -- Perform the second
  safeDivide result2 4              -- Perform the third
```

Here's how it works:

- In each line of the do block, a computation is performed within the monad.

- The result of each computation is extracted (if possible) and assigned to a variable using <-.

- The final result does **not** need <- (though it can if you want to bind it).

Under the hood, the above do block is equivalent to

```
exampleWithDo =
  safeDivide 10 2 >>= \result1 ->
  safeDivide result1 5 >>= \result2 ->
  safeDivide result2 4
```

Here's a basic structure of the "do notation" in any monad:

```
doSomething :: Monad m => m Int
doSomething = do
  x <- action1
  y <- action2
  action3 x y
  return (x + y)
```

155

CHAPTER 7   INPUT, TRANSFORMER STACK, OUTPUT

Behind the scenes, the do notation is translated into chains of >>= (bind) operations. The above example is equivalent to

```
doSomething =
  action1 >>= \x ->
  action2 >>= \y ->
  action3 x y >>
  return (x + y)
```

Let's look at some more examples using different monads:
List Monad:

```
combinations :: [Int] -> [Int] -> [Int]
combinations xs ys = do
  x <- xs
  y <- ys
  return (x + y)
```

This generates all possible sums of pairs from the two input lists.
IO Monad:

```
getUserInfo :: IO (String, Int)
getUserInfo = do
  putStrLn "What's your name?"
  name <- getLine
  putStrLn "How old are you?"
  age <- readLn
  return (name, age)
```

This interacts with the user to get their name and age.

Inside a do block, you can use "let" to create local bindings for pure computations without the need for in:

```
calculateBMI :: IO ()
calculateBMI = do
```

```
putStrLn "Enter your weight (kg):"
weight <- readLn
putStrLn "Enter your height (m):"
height <- readLn
let bmi = weight / (height ^ 2)
putStrLn $ "Your BMI is: " ++ show bmi
```

The rule of thumb when using the "do notation" is when you have a monadic computation, you have to use "<-" to bind it and when you have a pure computation, use "let."

You can also use pattern matching on the left-hand side of <-:

```
processData :: [(String, Int)] -> IO ()
processData dat = do
  (name, age) <- return (head dat)
  putStrLn $ name ++ " is " ++ show age ++ " years old."
```

DO NOTATION TO BINDINGS

Convert the "do notation" examples given above into the usual monadic chains using the >>= and >> operators.

By now, you should have a good understanding of the Functor–Applicative–Monad hierarchy as well as different types of monads and even different ways to write monadic code. Now we are finally ready to dive into structuring and writing real-world programs, and to do that we need to learn the last remaining piece of the puzzle.

CHAPTER 7   INPUT, TRANSFORMER STACK, OUTPUT

Monad Transformer Stacks

As I was learning Haskell, I was instantly captivated by the beauty of the mathematical concepts involved. We define types and functions, find similar abstractions, and make them instances of typeclasses such as Functor-Applicative-Monad. We use different types of monads for typical programming tasks - Reader for configuration/settings, Writer for logging, State for managing state in the pure fashion, lifting our pure computations into IO monad to avoid contamination, etc. - but there is one seemingly extremely big problem. How in the world do we combine all of these monads together?! Their purpose is to isolate and chain computations with each other, but *within a single monad*. We have a beautifully designed problem space that uses several. What to do?

Enter **Monad transformers**!

Monad transformers let you **stack** monads, letting you keep the neat composable structure of monads while juggling different effects.

A **monad transformer** is essentially a version of a monad that can operate "on top" of another monad, combining their powers. When you stack monads using transformers, you get the behavior of both at once, making it much easier to work with multiple effects in a single computation.

For instance

- You might need **stateful computations** that also **log** actions (StateT + WriterT).

- Or **reading from shared configuration** while handling **optional (Maybe) values** (ReaderT + MaybeT).

CHAPTER 7   INPUT, TRANSFORMER STACK, OUTPUT

Here's a quick overview of what monad transformers offer:

- Composition of effects: Combine multiple monadic effects into a single monad.
- Flexibility: Mix and match different monads based on your specific needs.
- Modularity: Add or remove effects without major code rewrites.
- Type safety: Maintain strong typing and the guarantees that come with it.

Each transformer has a corresponding "T" version of the monad you're familiar with: MaybeT, StateT, WriterT, ReaderT, etc. These transformers allow you to "lift" computations from one monad into another, so they can coexist peacefully and you can chain them together.

Let's continue the design of our "Blackjack" game, which

- Needs to keep track of the current **deck** (State)
- Needs to **log** all actions taken (Writer)
- Needs **configuration** settings like max players or game rules (Reader)

Dealing with these separately would be a nightmare of nested monadic operations:

```
type GameM = State Deck (Writer [String] (Reader Config a))
```

Without transformers, you would have deeply nested monads, and you'd need to constantly unwrap inner layers just to access your state, log, or configuration. Dwell on the above definition and imagine what it would take to get to the value of type "a."

Using monad transformers, you can **stack** these effects and manage them in a flat, readable way. You can then write clean code that deals with all of these effects as if they were one unified monad!

Now we can decode the State Monad definition from the previous chapter:

type State s = StateT s Identity

To do that, let us also look at the definition of StateT (see Figure 7-3):

newtype StateT s m a

Here "s" is our type for the State values and "m" is an underlying monad, which we are "transforming":

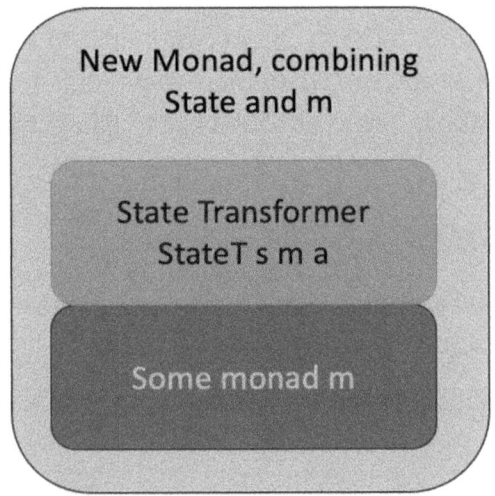

Figure 7-3. StateT transformer

Thus, the State monad definition itself is in fact a state transformer on top of a special monad called "Identity" (that does nothing – it's just another beautiful mathematical concept that we won't get into)! Sure, we could have defined the State monad separately from the transformer concept, but since we are virtually always stacking monads together,

CHAPTER 7   INPUT, TRANSFORMER STACK, OUTPUT

we might as well do it via a transformer as a more general abstraction and that's exactly what the *transformers* and *mtl* libraries offer. ReaderT, WriterT, MaybeT, etc. are all structured similarly and serve the same purpose – they turn any other monad into a combined monad that allows to mix effects of both.

Let's revisit our card game example and use **monad transformers** to stack the State and Writer monads. We'll keep track of the deck while logging the actions taken:

```
import Control.Monad.State
import Control.Monad.Writer

type GameM = StateT Deck (Writer [String])

-- Deal cards while logging the action
dealCardsT :: Int -> GameM [Card]
dealCardsT n = do
  deck <- get
  let (hand, newDeck) = splitAt n deck
  put newDeck
  tell ["Dealt " ++ show n ++ " cards."]
  return hand
```

Here, the StateT transformer manages the deck, while the Writer monad logs the actions. Notice that we can **combine** both monads seamlessly with do notation, and we don't need to painfully unwrap the log or state manually.

When stacking monads, you sometimes need to **lift** an operation from a lower-level monad into the current stack. The lift function helps you do just that:

```
lift :: (Monad m) => m a -> t m a
```

CHAPTER 7   INPUT, TRANSFORMER STACK, OUTPUT

For example, if you had a GameM that combined StateT and MaybeT, and you wanted to perform a Maybe operation inside your do block, you would **lift** it up into the monad stack:

```
dealOrFail :: MaybeT (StateT Deck IO) [Card]
dealOrFail = do
  hand <- lift $ dealCardsT 5   -- Lift the StateT action
                                   into MaybeT
  MaybeT $ return (Just hand)   -- Wrap the result in MaybeT
```

An example of a monad transformer stack for the final Blackjack game might look similar to the following:

```
type AppMonad = StateT GameState (ReaderT Config (WriterT Log IO))
```

This AppMonad combines state management (StateT), access to configuration (ReaderT), logging (WriterT), and IO operations, all in one powerful monad. The above is a pretty typical structure for real-world applications, even though you would also usually use some additional monad transformers from third-party libraries to work with databases, networking, etc.

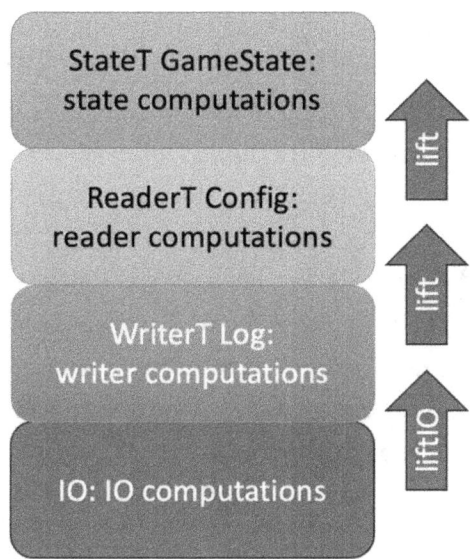

Figure 7-4. *Monad transformer stack for a typical real-world program*

Figure 7-4 may help you visualize the structure of your program. With time, this will become your second nature to separate your types and functions from the pure world and then "lifted" floors of different monads. You design types and pure functions first. Then you design monadic computations that live within each of your monads pretty much independent of each other, but keeping in mind that higher floors have access to the lower once, but not vice versa! Then if you need to perform an operation from a lower floor in the higher floor, you use "lift" (or several, and thus sometimes it is easier to define special operators that combine several lift functions together; otherwise, writing "lift $ lift $ liftIO" may become quite cumbersome). This might look a bit intimidating at first, but you will quickly start to appreciate the separation of concerns such a design gives you – and maintaining code written in this way becomes much easier as well, as you should only be concerned with every "floor" computation independent of each other.

Finally, let's discuss why IO Monad is always at the bottom of the transformer stack. It is Haskell's gateway to the "real world" of side effects – reading from files, writing to the console, making network requests, etc. In Haskell's pure functional world, **IO is sacred** because it represents computations that can have side effects, making them *impure* by definition.

As we discussed above, Haskell cleverly isolates impure computations inside the **IO** monad, ensuring that the rest of your code remains pure and referentially transparent. Once a computation crosses into the IO monad, there's no way back into the pure world. This is why you can only *run* an IO action as the **final result** of your program.

Key Property: IO is the endpoint. You can't escape from IO.

Monad transformers allow us to stack different effects together so that we can combine state management, logging, configuration reading, and more *alongside* IO. However, the **IO** monad must always remain at the **bottom** because

- IO is the most "real-world" monad: All other monads represent computational effects that are logically contained *within* Haskell's pure framework. Things like State, Reader, and Writer are purely functional abstractions – IO is the only monad responsible for interacting with the external, imperative world.

- IO is impure: As soon as you introduce IO into a computation, you're essentially saying, *"This computation may have side effects."* Once you've reached this point, you can't go back to a pure monad without side effects. The non-IO layers of your transformer stack are **pure abstractions** and can only exist **outside** IO.

CHAPTER 7   INPUT, TRANSFORMER STACK, OUTPUT

When you stack multiple monads together using monad transformers, you sometimes need to perform actions from one of the lower monads (like IO) while inside a transformer layer (like StateT or ReaderT). To do this, you use the lift function described above to "pull up" an action from the underlying monad or "lower floors."

For another example, here's a transformer stack that combines StateT with IO:

```
type AppM = StateT GameState IO

runGame :: AppM ()
runGame = do
  liftIO $ putStrLn "Welcome to the game!"  -- liftIO lifts an
  IO action into the stack
  gameState <- get
  -- Rest of the game logic...
  return ()
```

In this case

- **liftIO** brings an IO action into the StateT layer.
- **Why do we lift?** Because the IO monad is at the bottom of the stack, and we need to explicitly lift it up into the current transformer layer.

If **IO** was not at the bottom, we wouldn't be able to lift it – the whole monad stack would lose its ability to interact with the outside world!

What happens if we tried to place IO *higher up* in the monad stack, such as something like this?

```
type BrokenMonadT = IO (StateT GameState (WriterT [String]
Identity))
```

This is **not allowed** in Haskell! The reason is that the **ordering of effects matters**:

- If IO is on top, you could theoretically "escape" from an IO action back into a pure computation (like State). This would break Haskell's guarantees about separating pure and impure code.
- Purity comes first: The other layers (like State, Writer, or Reader) operate in a pure, referentially transparent way *inside* the monad transformer stack. IO, however, breaks this purity by introducing side effects – so it needs to be encapsulated and managed as the final effect.

By keeping IO **at the bottom**, we ensure that any side effects are structurally contained, and the pure functional layers remain untouched by the impurity until the very end.

Finally, Monad transformers are composable, but IO isn't. Most monads, like State, Writer, or Reader, can be easily composed and unwrapped. You can take a pure State computation and transform it back into a pure value and state:

```
runState :: State s a -> s -> (a, s)
```

But this is **not possible with IO**. Once you've entered the world of IO, there's no escape:

```
runIO :: IO a -> a
```

CHAPTER 7   INPUT, TRANSFORMER STACK, OUTPUT

This function does **not exist** in Haskell! You can't simply convert an IO a computation into a pure value a without running it. This is why IO must always be at the bottom – other monads can be composed and peeled off, but IO is the endpoint of the stack.

Let's put everything together with a simple code example that combines **stateful computations** (StateT) with **IO operations** like reading from the console:

```haskell
import Control.Monad.State

-- Define a simple game state
type GameState = Int

-- Define a monad stack that combines state and IO
type GameM = StateT GameState IO

-- A function that modifies the game state and interacts with the user
playGame :: GameM ()
playGame = do
  liftIO $ putStrLn "Guess the number:"
  guess <- liftIO getLine
  let num = read guess :: Int
  put num    -- Update the state with the player's guess
  liftIO $ putStrLn $ "Your guess was: " ++ show num

-- Run the game by providing an initial state and executing the IO at the bottom
main :: IO ()
main = evalStateT playGame 0    -- `evalStateT` runs the StateT monad, leaving IO as the final effect
```

CHAPTER 7 INPUT, TRANSFORMER STACK, OUTPUT

Here

- **GameM** is a transformer stack that combines StateT on top of IO.
- Inside the computation, we use **liftIO** to bring IO actions (like putStrLn and getLine) into the StateT monad.
- The program keeps IO at the bottom of the stack, ensuring that all side effects occur only at the very end.

IO always stays at the bottom.

To sum it up, **IO must always be at the bottom of your monad transformer stack** for the following reasons:

- IO represents side effects: Once you're in the IO monad, you can no longer return to the pure world. This makes it a natural endpoint.
- Purity is preserved: Monad transformers like StateT, ReaderT, and WriterT work in the pure functional realm and must remain on top of IO to keep pure computations intact.
- Lifting actions: You can "lift" actions from IO up into the transformer stack, but you can't move pure computations back down into IO.

Keeping IO at the bottom ensures that you maintain Haskell's purity guarantees while still being able to interact with the world through side effects – neatly encapsulated and controlled.

While monad transformers can be complex, they provide a structured way to handle multiple effects in your Haskell programs. They're like the ultimate spell-combining technique for our wizard analogy – allowing you to weave together different types of magic into one cohesive and powerful spell!

Conclusion

Now we've learned enough theory and concepts, and it is finally the time to put all our concepts to practical use and start writing actual real-world Haskell programs! In this chapter, we learned how to perform basic input and output using IO monad and built our first Monad transformer stacks – the primary way to structure Haskell programs. We will be building several different variants of these in the next chapters, so buckle up!

CHAPTER 8

Blackjack: Full Haskell Program

Now we are ready to apply everything we've learned and design and then build the Blackjack program from scratch to win the Haskell magician competition. This is the only chapter where we will carefully discuss step by step building a program from nothing to fully functional, down to the "main" function. It should allow you to follow the logic and then apply it when building your own, much more complicated projects, as well as follow the more advanced concepts we will be reviewing in the next chapters.

Let us recall the simplified Blackjack rules that we decided to implement:

- The dealer deals two cards to each player and one card to themselves.
- Players make bets (we will not allow any splits and others to simplify the game).
- Players take cards until they say "stop" or go bust (over 21 total score).
- The dealer takes cards until they hit 17 or go bust.
- The score is calculated as 10 for J, Q, K, either 1 or 11 for Ace, and at face value for the other cards.

- If the player is closer to 21 than the dealer, they get paid their bet; otherwise, they lose it.

- If the player gets 21 from the initial dealing, they win 2× their bet.

Another simplification we will make is that we will make our game single player. Now we will forget somewhat painful, but very educational, exercises that we did for the Cards domain when we "invented" the State monad for the DeckM type and will rather use all the power of existing Haskell libraries. Let us recap and apply the process to solve problems in the functional way:

- Design the problem domain in terms of types. We did most of this work in the previous chapters. Now we'll just need to expand it to include types that will help us handle state and input/output – and thanks to the previous chapter, we now can.

- Think how we will structure the whole program in terms of the monad transformer stack.[1] As discussed in the last chapter, IO will always be at the bottom, and then the usual choice is Reader, Writer, and State to manage tasks related to configuration, logging, and state. In many real-world tasks, you will also need some way to handle errors – a separate topic as there are different approaches to deal with this in Haskell, but adding another monad to the stack for error handling is quite common as well. For our simplified Blackjack program, we will only use State on top of IO – adding

---

[1] Monad transformer stacks are not the only possible abstraction to structure Haskell programs, but certainly are the most common. It is always good to start with it and then move to more esoteric approaches, for instance, possible with Arrows.

new layers adds nothing new in terms of the concepts learned, and you'll easily expand this approach to more complex stacks. We will also have examples of those in the next chapters.

- Design additional types and functions that live on corresponding "floors" of our monad transformer stack. Design additional interface functions on top of the standard typeclass functions if needed (e.g., if we have several monads in the stack – providing functions that combine "lifts" of different sizes may prove handy).

- Then everything that's left is connecting our functions – "spells" – so that the types compose. And voila! Our program is ready!

PREPARATION EXERCISE

Review the types and functions we have designed for the Cards domain in the previous chapters. Think about what is lacking to handle the state of the game and what input/output functions you are going to need. Separate them into different monad stack "floors."

CHAPTER 8   BLACKJACK: FULL HASKELL PROGRAM

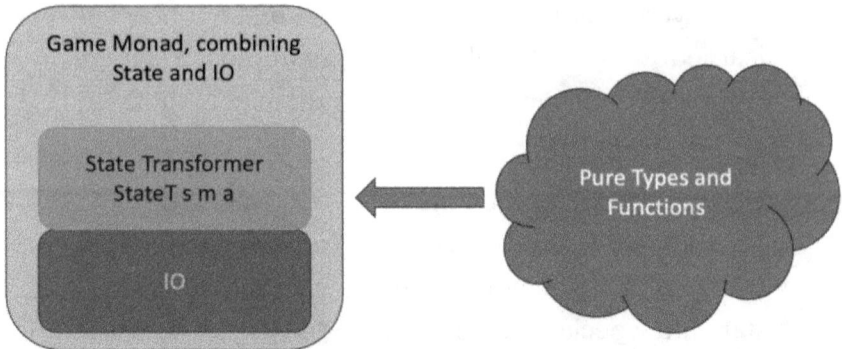

Figure 8-1. *Thinking about program design. Pure types and functions describe our problem domain as much as possible. They are then used inside our monad stack as needed to perform computations with mathematically guaranteed results, while side effects of changing state and interacting with the user are hidden inside the respective monads and will never pollute our pure world*

Let's roll up our sleeves and get to work!

Preparation

By now you should have installed Haskell from Haskell.org – the specific ways to do that change with time, but the website should remain. Haskell uses tools called `cabal` and `stack` for package management and for developing your projects. We will not be discussing in detail their usage; they both have great extensive documentation, and I encourage you to review it. Stack is enough for most practical purposes, so I suggest you start there. Another necessary resource for Haskell development is hackage.haskell.org – it contains *all* libraries that you will use in your practical development. So whenever I reference libraries such as *transformers* or *mtl*, go to Hackage and search for them to read the detailed documentation.

CHAPTER 8   BLACKJACK: FULL HASKELL PROGRAM

The code that we will be discussing in this chapter is in the book's GitHub repository in the folder for Chapter 8.

To start the new Haskell project, go to your development folder and run stack new <project_name>, which will create the new folder and initialize some default settings. Normally, the directory structure will be similar to the one shown in Figure 8-2. You can use your favorite coding editor for Haskell development – VS Code, Atom, or even vi or emacs if you are into terminals (but then you probably don't need to read this book to be fair).

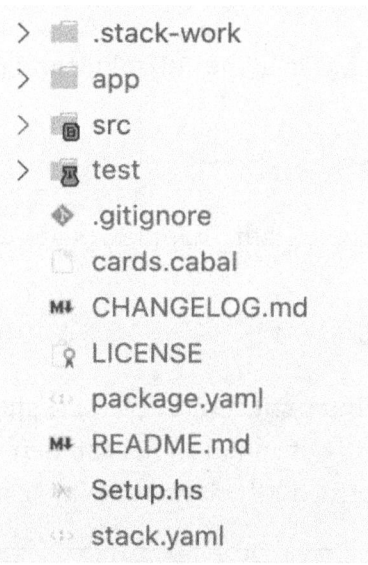

Figure 8-2. *Typical Haskell project structure*

Usually, we put all the "libraries" code in the **src** folder, while **app** is the "default" executable folder that will contain the file with our main function code. We can have several executables in one project as we will see down the line. Go to package.yaml and review it – it contains various project settings, which are pretty self-explanatory. For now, simply change the executable name to "blackjack" as shown below and keep the rest intact:

```
...
library:
  source-dirs: src
executables:
  blackjack:
    main:              Main.hs
    source-dirs:       app
    ghc-options:
```

Now, edit the Main.hs file in the app folder and put a placeholder main function there:

```
main = putStrLn "Hello World"
```

After that, in the terminal run a couple of stack commands:

```
stack build
stack exec blackjack
```

This should build your project (may take a slightly long time the first time as stack will need to download all needed dependencies and build them as well) and then run it, which will display "Hello World".

There is one important setting in stack.yaml file called "resolver." You can keep it as is if you are just starting out, but it defines all the versions of all the libraries and GHC – so pick carefully in real projects. Usually some "lts" setting is the best, as it will set up dependencies for long-term supported libraries.

Now that we have a basic project structure setup, we are ready to design and code.

Initial "Pure" Design

First, we should always start with thinking about our problem domain and how it can be efficiently represented using pure types and functions. We have been discussing a lot with regard to how it can be done for our Blackjack game in the previous part of the book, so it should be easy for you to follow. Let's create a file called Cards.hs in the **src** folder and put all the pure types and functions for the Cards problem domain that we have designed in the early chapters:

```haskell
{-#LANGUAGE OverloadedStrings #-}
module Cards
where

-- simple algebraic data types for card values and suites
data CardValue = Two | Three | Four | Five | Six | Seven | Eight |
  Nine | Ten | Jack | Queen | King | Ace
  deriving (Eq, Enum)
data CardSuite = Clubs | Spades | Diamonds | Hearts deriving
  (Eq, Enum)

-- our card type - merely combining CardValue and CardSuite
data Card = Card CardValue CardSuite deriving(Eq)

-- synonym for list of cards to store decks
type Deck = [Card]

instance Show CardValue where
  show c = ["2", "3", "4", "5", "6", "7", "8", "9", "10", "J",
    "Q", "K", "A"] !! (fromEnum c)
```

CHAPTER 8   BLACKJACK: FULL HASKELL PROGRAM

```haskell
instance Show CardSuite where
  show Spades   = "♠"
  show Clubs    = "♣"
  show Diamonds = "♦"
  show Hearts   = "♥"

-- defining show function that is a little nicer then default
instance Show Card where
  show (Card a b) = show a ++ show b

-- defining full deck of cards via comprehension; how cool is that?! :)
fullDeck :: Deck
fullDeck = [ Card x y | y <- [Clubs .. Hearts], x <- [Two .. Ace] ]

smallDeck :: Deck
smallDeck = [Card Ace Spades, Card Two Clubs, Card Jack Hearts]

dealCards :: Int -> Deck -> (Deck, Deck)
dealCards n deck = let newDeck = drop n deck
                       plDeck  = take n deck
                   in (newDeck, plDeck)

countCards :: Deck -> (Int, Deck)
countCards deck = (length deck, deck)
```

Here we have the types defined for representing our cards, a handy typeclass instance for **Show** typeclass so that our cards look nicer in the terminal output, and a couple of handy functions that deal and count the cards.

The {-#LANGUAGE OverloadedStrings #-} line at the top of our file defines various GHC options. GHC stands for "Glasgow Haskell Compiler" and is the most powerful and most developed Haskell compiler, even though there are others. It has many, many various extensions that come into the language directly from research papers, and many of them allow us to do some amazing stuff with types, but unfortunately we will only touch upon this topic very lightly – perhaps in more detail in a future book, as it deserves a book of its own. The "OverloadedStrings" option is the one you'll likely use most often – it allows you to write code like

```
myString :: Text
myString = "Hello World"
```

Normally, it would not be allowed as literal strings in Haskell code must be of type **String**, which is just a synonym to very straightforward **[Char]** – list of chars – so you would have to provide explicit conversion from **String** to **Text** that gets tedious to write. **Text** on the other hand is a much more efficient implementation of packed strings with full Unicode support as well as other advantages – so whenever you want to work with text efficiently, you want to use **Text**. Unfortunately, a lot of useful libraries still use **String** for some functions, so you have to pay attention when writing production code. We'll discuss this in more detail in the next chapters – especially the ones that deal with the Web.

The next lines define the name of our module – Cards – and eventually the list of imports, which is currently empty:

```
module Cards
where
```

The name of the module has to be the same as the file name, and it also follows the directory structure – if we created a folder called "Game" inside src and put "Cards.hs" there, the name of the module would have to be Game.Cards. Very straightforward to follow.

CHAPTER 8 BLACKJACK: FULL HASKELL PROGRAM

You can also test your current types and functions by loading your project into the interactive ghci environment using another handy stack command:

```
stack ghci
...
Loaded GHCi configuration from ...
ghci> let (deck, myDeck) = dealCards 5 fullDeck
ghci> myDeck
[2♣,3♣,4♣,5♣,6♣]
ghci>
```

This command loads all the modules of your project with all necessary dependencies, which means you can interactively test all of your types and functions – extremely handy to experiment and find potential issues with your code and absolutely necessary when learning Haskell. A couple of very useful ghci commands are ":i" that displays information on pretty much anything in your program and ":t" that gives you type information:

```
ghci> :i String
type String :: *
type String = [Char]
        -- Defined in 'GHC.Base'
ghci> :t myDeck
myDeck :: Deck
ghci> :i Deck
type Deck :: *
type Deck = [Card]
        -- Defined at .../ch08/src/Cards.hs:14:1
```

Feel free to explore a bit.

CHAPTER 8 BLACKJACK: FULL HASKELL PROGRAM

Now, what is missing from our small pure world of the Cards problem domain for our simplified Blackjack game? Of course, this is some data types to keep track of the "global" state for the player, which in our game is simply the amount of money, as well as the "local" state for hands for both the player and the dealer.

> **EXERCISE**
>
> Design these types and related useful functions and extend Cards.hs, respectively. For instance, calculating the Blackjack score of the hand function is currently missing!

Designing the types for the states as well as functions that calculate card and hand values is relatively straightforward, so let's add the following code to our Cards.hs file. Some explanations are given in the comments, so read the listing carefully:

```
-- another type synonym for list of cards to provide additional
clarity in the types below
type Hand = [Card]

-- state for the player
data Player = Player {
    playerHand :: Hand
  , playerBet :: Int
  , playerMoney :: Int } deriving Show

-- state for the dealer
data Dealer = Dealer { dealerHand :: Hand } deriving Show

-- convenience functions that extract rank and suit from
any Card
rank :: Card -> CardValue
```

CHAPTER 8   BLACKJACK: FULL HASKELL PROGRAM

```haskell
rank (Card r _) = r
suit :: Card -> CardSuite
suit (Card _ s) = s

-- Calculate the blackjack value of card. Ace is tricky!
cardValue :: Card -> Int
cardValue (Card rank _) = case rank of
    Ace -> 11   -- We'll handle Ace's 1 or 11 value in handValue
    Jack -> 10
    Queen -> 10
    King -> 10
    _ -> fromEnum rank - fromEnum Two + 2

-- Calculate the value of a hand. This function is not entirely
correct, pay attention :)
handValue :: Hand -> Int
handValue hand =
    let baseValue = sum $ map cardValue hand
        aces = length $ filter (\c -> rank c == Ace) hand
    in if baseValue > 21 && aces > 0
       then sum $ map (\c -> if rank c == Ace then 1 else
       cardValue c) hand
       else baseValue
```

The function that calculates the hand value – handValue – is a bit tricky, but with the knowledge you have by now, you should decipher it easily. First, we are calculating baseValue by mapping over all cards in the hand and getting their face values. Then, if the resulting value is over 21 and we have Aces in the hand, we recalculate by making *all* Ace values equal to 1. Do you see where the problem is?

CHAPTER 8   BLACKJACK: FULL HASKELL PROGRAM

> **FIX THE ACES PROBLEM**
>
> Fix the handValue function so that it calculates the value of multiple Aces correctly. Alternatively – and if you are really into cards – expand the final program to give the player the ability to "split" their hand if they get two Aces.

Load the new code into ghci with stack ghci program and test the new functions.

The only missing piece in our pure Cards world is the overall state of the game that should combine all our "smaller" states. Here it is:

```haskell
data GameState = GameState
    { player :: Player
    , dealer :: Dealer
    , deck :: Deck
    } deriving Show
```

This state type captures everything we need to keep track of in our game: player state, dealer state, and the state of the game deck. We could have easily made a game deck part of the Dealer since it is kind of a dealer's deck, but this is a free design choice – you can do it differently just as easily.

The above gives us a solid pure foundation, and now we can move to constructing our monad stack so that we are able to handle the effects – randomizing decks, changing state, handling input/output!

Building the First Floor: IO

We have agreed that we will use a very simple monad stack for our program – **StateT** on top of **IO**. The goal of this chapter is to learn the principles, and expanding them to more layers or "floors" is very straightforward. But before we move to actual code, let us once again

CHAPTER 8   BLACKJACK: FULL HASKELL PROGRAM

review Figure 8-1 that shows the high-level architecture of our program. We have a pure world that deals with modeling the problem domain and most of the internal business logic and a stack of two monads: State to update the game's state and IO to interact with the player. Our next task is to continue designing types and functions, now related to our monad stack, but a very important step in this process is *deciding which functions should live on which floors.*

Let's take a function of shuffling the deck as an example. We cannot implement it in the pure world, because it involves randomness, and randomness is provided in the IO monad. However, since our "final" program monad will be StateT on top of IO, we may want to run lots of functions, including shuffling the deck, directly there. If you recall in the last chapter, we have "lift" and "liftIO" functions that can help us with that. Also recall that functions from the lower floors of the monad stack can be lifted to the higher floors, but *functions from the higher floors cannot be brought down to the lower floors.* This gives us an important principle:

If a function needs to work on several "floors" of your monad transformer stack, always put it to the lowest possible level. That is, if a function can be pure, make it pure – that's the most important principle – and then you can use it anywhere. If a function needs IO, but doesn't need State, put it into IO. And so on.

With that, let's start building our IO floor. Create a file called IOFloor.hs (of course this name is somewhat cheesy – feel free to change it to whatever suits your architectural design tastes best), and let's put this initial code there:

{-#LANGUAGE OverloadedStrings #-}

module IOFloor
where

CHAPTER 8   BLACKJACK: FULL HASKELL PROGRAM

```
import System.Random
import Data.List (sortOn)
import Cards

shuffleDeck :: Deck -> IO Deck
shuffleDeck deck = do
    gen <- newStdGen
    return $ map snd $ sortOn fst $ zip (randoms gen ::
    [Int]) deck
```

In order for it to compile when you run stack build, you will need to add an explicit dependency on random package to package.yaml:

```
dependencies:
- base >= 4.7 && < 5
- random
```

This is the usual pattern if you want to use an external library – you import what you need in the import statements of your source code and then add a dependency to the package to package.yaml. In some cases you may need to give some additional info, such as versions, or reference to an external source and others, but stack is usually very helpful in providing messages that guide you through that process. In this book further on, let's focus on code.

Another notable thing in this code is that it follows our design picture quite closely – we import the Cards module, which defines our pure functions and types, to use on the IO floor. But Cards itself does not import anything that has even the slightest thing to do with the IO, thus remaining independent and pure.

Once your code compiles, run stack ghci and test that our shuffling function works:

```
ghci> d <- shuffleDeck fullDeck
ghci> d
```

CHAPTER 8   BLACKJACK: FULL HASKELL PROGRAM

[4♠,A♣,3♣,A♥,5♠,4♣,Q♥,10♣,2♣,Q♠,7♠,8♠,5♣,5♥,J♦,10♥,5♦,
4♥,3♠,2♦,6♠,2♥,J♠,4♦,8♦,K♥,7♣,A♠,9♠,K♠,2♠,6♣,Q♦,9♣,Q♣,
8♥,J♥,K♣,6♦,7♥,9♥,10♦,3♦,A♦,3♥,10♠,K♦,6♥,9♦,7♦,J♣,8♣]
ghci>

Now, why did we write `d <- shuffleDeck fullDeck` instead of `let d = ...`? Think and answer before reading further.

If you look at the type of our shuffling function, it says `Deck -> IO Deck`. This means it is a computation in the IO monad that "returns" the value of type Deck, but lifted in the IO monad. And if you read the last chapter carefully, you know that there is no escaping the IO monad, but instead using so-called do notation, we can bind pure values to variables from IO computations – and this is exactly what we do here. Now that we have a "pure" Deck bound to the "d" variable, we can further inspect it inside ghci, for instance, by running `handValue d` to get 340 as the answer. Without the do notation we could have written something like `shuffleDeck fullDeck >>= pure . handValue` with the same effect. Dwell on it and re-read the last chapter if needed. These things may be confusing at first, but if you carefully look at the types, everything should become clear.

But why did we make our shuffleDeck function an IO computation in the first place? To answer that, go to hackage.haskell.org, search for System.Random, and let's read the docs for `newStdGen` (Figure 8-3).

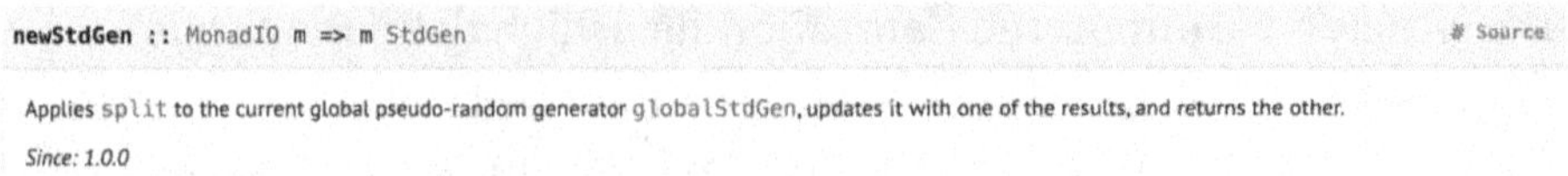

```
newStdGen :: MonadIO m => m StdGen                                          # Source
```
Applies `split` to the current global pseudo-random generator `globalStdGen`, updates it with one of the results, and returns the other.

Since: 1.0.0

Figure 8-3. *Documentation for newStdGen*

The type signature tells us that it's a monadic action in some monad m that is part of the MonadIO typeclass. The latter is a very handy method to generalize library methods to custom monads using IO under the hood, but we won't get into it now. What we need to know is that IO itself is indeed part of MonadIO. So the only way for us to work with random numbers (apart from writing a pure pseudo-random generator yourself – might be possible, but not very practical as it will be fully deterministic by definition) is by being inside IO. Hence, we *have to* make the shuffling function an IO computation. Haskell's type system forces us to make certain decisions, but these decisions protect the purity and guarantees of the language.

I leave decoding of what the function actually does to the reader – it's a good exercise that you can easily do by now. The only missing piece there is the `randoms` function, which is also part of System.Random – so read on it while you are on Hackage. To give you a hint, `zip` ... creates a list of tuples, where each tuple contains a random integer and a card from the deck. The rest is just some basic list manipulation.

What other functions do we need on our IO floor? Of course, those are functions that get input from the user and show the current state of the hands and money to the user.

EXERCISE

Design and write the remaining functions that should live on the IO "floor" of our monad stack.

Once you are done, you can check how we have implemented it in the repository code, and let's move to the top floor of our stack – the actual Game monad, built as StateT on top of IO.

CHAPTER 8   BLACKJACK: FULL HASKELL PROGRAM

Building the Top Floor: StateT IO

Create another file called GameFloor.hs in the src folder, and let's build the top floor for the monad stack of our game:

```
{-#LANGUAGE OverloadedStrings #-}
```

```
module GameFloor
where
```

```
import Control.Monad.State
import Cards
```

```
-- Our monad transformer stack
type Game = StateT GameState IO
```

For this to compile, you need to carefully write all imports and then add mtl to the list of dependencies in package.yaml just as we did with random previously. We are using the **mtl** library here, even though we could have used **transformers**. Feel free to familiarize yourself with both, but the important thing to keep in mind is that mtl is built on top of transformers and provides some additional handy features.

The most important line in the code above is of course the definition of the Game type: `type Game = StateT GameState IO`. It tells us that Game is a State monad transformer with state represented by GameState, which we defined above, and living on top of the IO monad. It literally recreates the scheme in Figure 8-1 in the very beginning of this chapter. **Game** is also a monad itself, as any monad transformer stack, so we can apply everything we learned about monads and the State monad in particular in the previous chapters.

For instance, it means that inside any **Game** computation, we can write `get >>=` or `state <- get` (using do notation) and get access to the current state (that will be of type GameState), while we can also use `liftIO` to run *any* IO computation. That's basically it. That's pretty much all we need to know to finish our program! But of course I encourage you to read the docs for the State monad in the mtl library, as usual – learning key libraries is essential to becoming a high-performing Haskell programmer. They are like the spell books for magicians! The more you know, the more powerful you become.

Now we can design the actual game logic, using pure types and functions from Cards and IO types and functions from the IO floor of our stack. The game goes as follows:

- The player bets.

- The dealer deals two cards to the player and one to themselves.

- We show all cards to the player.

- The player either "hits" – and is dealt another card until they go bust – or says "stop."

- The dealer takes cards until they either hit 17 or more or go bust.

- We adjust the player's balance according to who won.

BUILD THE GAME FUNCTIONS

Build the computations in the Game monad that implement the functionality described above. As you are working on this, you may have ideas on how to expand the pure Cards.hs world to make your life easier as a programmer building the rest of the game. One example can be a function that modifies the player status in such a way that we either add or take some money to or from their account. What other examples are there? Expand the pure world – and remember the principle: if a function *can* be made pure, it should.

CHAPTER 8 BLACKJACK: FULL HASKELL PROGRAM

Let us build all these functions one by one, expanding our pure and IO worlds and floors as we go. The following is a very detailed walk-through of an actual typical design approach to writing Haskell programs – we will not be as detailed in the next chapters of this book, but the principles learned here will help you fill the gaps in more advanced concepts.

Start of the round: a player bets. We need to ask the player how much, decrease their account correspondingly, and proceed further. We can implement this function fully inside Game, but we remember the principle discussed above – the computations should be put as low in the stack as possible. Applying this principle together with the idea of decomposing the problem into easier pieces, we decide that a much better design is to have two functions: a first in IO that asks the player for their bet and returns a number and a second in Game that adjusts the player's status, respectively. Further thinking through this, we decide to add a third, pure function, makeBet, that makes code much more readable. This design is cleaner, easier to maintain, and more general!

Here are the code changes:

```
-- IOFloor.hs
...
import Text.Read (readMaybe)
...
-- get a bet number from the player
getPlayerBet :: IO Int
getPlayerBet = do
    bet <- do
        putStrLn "Enter your bet (min: 10, max: 100):"
        readMaybe <$> getLine
    case bet of
        Just b | b >= 10 && b <= 100 ->
            pure b
        _ -> do
```

CHAPTER 8 BLACKJACK: FULL HASKELL PROGRAM

```
            putStrLn "Invalid bet. Please try again."
            getPlayerBet
-- Cards.hs:
-- convenience pure function modifying player state after
making a bet:
makeBet :: Int -> Player -> Player
makeBet bet pl = pl { playerBet = bet, playerMoney =
(playerMoney pl) - bet}

-- GameFloor.hs:
...
-- initial game state
initialState :: GameState
initialState = GameState
    { player = Player [] 0 1000
    , dealer = Dealer []
    , deck = fullDeck
    }

-- computation in Game that asks a player for a bet and makes
needed state adjustments
playerBetAction :: Game ()
playerBetAction = do
    bet <- liftIO getPlayerBet
    pl <- gets player
    let pl' = makeBet bet pl
    modify (\s -> s { player = pl'})
```

Some comments to the above. The getPlayerBet IO function can be improved by using a function such as maybe that we briefly mentioned in the early chapters, but the current code even though a bit clumsy is easier to read. Also, you may note that min and max bets are now hardcoded – a bad design – and an obvious improvement is to add some read-only game

CHAPTER 8   BLACKJACK: FULL HASKELL PROGRAM

settings and manage them by using which abstraction? Of course, the Reader monad! That will be a nice exercise to improve the current game, won't it?

The first computation from the Game floor – playerBetAction – uses getPlayerBet to get a number from the player and then makes necessary adjustments to the state. It uses everything we've discussed so far about state transformers – lifting a function from IO to our higher floor with liftIO, getting the current state via gets, adjusting it with the help of a handy pure function, and then modifying the game state with the help of modify (read the documentation for the State monad from the mtl library to understand what other functions are available), which puts a new player state into our GameState. Very concise code for so many things happening!

You may also notice that we don't check whether the user is betting more than they have left. You guessed it right – another good exercise ☺

As usual, you can test the functions we just added in ghci by running stack ghci and then

```
ghci> execStateT playerBetAction initialState
Enter your bet (min: 10, max: 100):
300
Invalid bet. Please try again.
Enter your bet (min: 10, max: 100):
50
GameState {player = Player {playerHand = [], playerBet = 50,
playerMoney = 950}, dealer = Dealer {dealerHand = []}, deck =
[2♣,3♣,4♣,5♣,6♣,7♣,8♣,9♣,10♣,J♣,Q♣,K♣,A♣,2♠,3♠,4♠,5♠,6♠,
7♠,8♠,9♠,10♠,J♠,Q♠,K♠,A♠,2♦,3♦,4♦,5♦,6♦,7♦,8♦,9♦,10♦,J♦,
Q♦,K♦,A♦,2♥,3♥,4♥,5♥,6♥,7♥,8♥,9♥,10♥,J♥,Q♥,K♥,A♥]}
ghci>
```

CHAPTER 8   BLACKJACK: FULL HASKELL PROGRAM

execStateT is another function from the State monad standard interface – it takes a StateT action (which in our case is a Game action) and initial state, runs the action, and returns final state. As you can see, our code runs correctly – a 50 bet is being correctly shown in the final state.

Let's continue building our Game floor!

The dealer deals two cards to the player and one to themselves. To deal n cards to the player, we have to do a bunch of different GameState manipulations that are very straightforward:

```
dealCardsToPlayer :: Int -> Game ()
dealCardsToPlayer n = do
    -- read current GameState
    gs <- get
    let pl = player gs
    -- deal cards using pure function
    let (newDeck, playerDealt) = dealCards n (deck gs)
    -- create new player status with new cards
    let pl' = pl { playerHand = (playerHand pl) ++ playerDealt }
    -- update game status
    modify (\s -> s { player = pl', deck = newDeck})
```

This should be easy to follow. We are using the pure functions we have already defined and then updating different states step by step. This code can be optimized quite a bit, plus a function that deals to the dealer repeats a lot of it, so making the two computations more concise is another nice exercise for you.

Show all (currently dealt) cards to the player. Again, also very straightforward:

```
showPlayerState :: Game ()
showPlayerState = do
    pl <- gets player
    liftIO $ print pl
```

CHAPTER 8   BLACKJACK: FULL HASKELL PROGRAM

Same for the dealer's hand.

Rounds of the player hitting or stopping, then the dealer takes his cards, and then we determine the round winner. This functionality combines what we've defined before into some sort of a round loop:

```
-- GameFloor.hs:
-- part when the dealer takes cards
-- we are passing the final player hand score to compare
dealerAction :: Int -> Game ()
dealerAction pvalue = do
    dl <- gets dealer
    let value = handValue (dealerHand dl)
    if value < 17
        then do
            dealCardsToDealer 1
            showDealerHand
            dealerAction pvalue
        else do
            showDealerHand
            liftIO $ putStrLn $ "Dealer stands with " ++
            show value
            if ((value > 21) || (value < pvalue))
                then do
                    liftIO $ putStrLn "Player wins!"
                    pl <- gets player
                    let pl' = payout ((playerBet pl)*3) pl
                    modify (\s -> s {player = pl'})
                    showPlayerState
                else do
                    liftIO $ putStrLn "Player looses :("
                    pl <- gets player
                    let pl' = payout 0 pl
```

```haskell
            modify (\s -> s {player = pl'})
            showPlayerState

-- main player round
playerAction :: Game ()
playerAction = do
    showDealerHand
    showPlayerState
    action <- liftIO $ do
        putStrLn "Do you want to (H)it or (S)tand?"
        getLine
    case action of
        "H" -> do
            dealCardsToPlayer 1
            pl <- gets player
            let newValue = handValue (playerHand pl)
            if newValue > 21
                then do
                    liftIO $ putStrLn "You bust!"
                    showPlayerState
                    let pl' = payout 0 pl
                    modify (\s -> s {player = pl'})
                    showPlayerState
                    -- modifyPlayer player (\p -> p {
                        playerMoney = playerMoney p -
                        playerBet p })
                else if newValue == 21
                    then do
                        liftIO $ putStrLn "21! Paying out 2x!"
                        showPlayerState
                        let pl' = payout ((playerBet pl)*3) pl
                        modify (\s -> s {player = pl'})
```

```
                    showPlayerState
                 else playerAction
    "S" -> do
        showPlayerState
        liftIO $ putStrLn "You stand"
        pl <- gets player
        let newValue = handValue (playerHand pl)
        dealerAction newValue
    _ -> do
        liftIO $ putStrLn "Invalid action. Please choose
        H or S."
        playerAction

-- Cards.hs:
-- another helper function
payout :: Int -> Player -> Player
payout amount pl = pl { playerBet = 0, playerMoney =
(playerMoney pl) + amount, playerHand = [] }
```

The code above is purposely made very verbose and straightforward. There is a lot of room for optimization using usual Haskell techniques: you can add several helper functions, both pure and in the Game monad, such as modifying the player's state within the Game monad; you can optimize the branching using guards and similar techniques; etc. Try that before we finish the program.

EXERCISE

Optimize the playerAction and dealerAction methods above.

CHAPTER 8 BLACKJACK: FULL HASKELL PROGRAM

All that is left by now is to implement the main game cycle, where we shuffle the deck between rounds, ask a player for the bet, and then run the player round.

Final Game and Recap

The final computation that is left to add is the main game cycle:

```
gameCycle :: Game ()
gameCycle = do
    -- asking for a bet first
    playerBetAction
    d <- gets deck
    -- checking if the deck has fewer than 15 cards left
    if (length d) < 15
       then do
          -- if yes, resetting to full deck
          modify (\s -> s {deck = fullDeck})
          gameCycle
       else do
           -- shuffling the deck first
           d' <- liftIO $ shuffleDeck d
           modify (\s -> s {deck = d'})
            -- dealing cards
           dealCardsToPlayer 2
           dealCardsToDealer 1
            -- cycling game rounds
           playerAction
            -- resetting the dealer's hand
           dl <- gets dealer
           modify (\s -> s { dealer = dl {dealerHand = []}})
           gameCycle
```

197

CHAPTER 8   BLACKJACK: FULL HASKELL PROGRAM

The code above is also prone to a couple of small optimizations, such as moving the dealer hand reset code to the main round cycle and others. In any case, by now we have fully finished building the Game floor, which is "just" a StateT GameState IO monad transformer stack. The only thing that's left is to define our "main" function in the Main.hs file:

```
{-# LANGUAGE OverloadedStrings #-}

module Main where

import Control.Monad.State
import GameFloor

main = execStateT gameCycle initialState
```

Our main computation is extremely simple: we run the gameCycle action with initialState with the help of the execStateT function from the State monad library interface that we discussed above. That's it. We are done! You can build the program, run stack exec blackjack, and test that everything works:

```
{1:59}~/dev/magicalhaskell/ch08:main X ⇨ stack exec blackjack
Enter your bet (min: 10, max: 100):
100
Dealer {dealerHand = [J♠]}
Player {playerHand = [5♣,Q♦], playerBet = 100,
playerMoney = 900}
Do you want to (H)it or (S)tand?
S
Player {playerHand = [5♣,Q♦], playerBet = 100,
playerMoney = 900}
You stand
Dealer {dealerHand = [J♠,7♦]}
Dealer {dealerHand = [J♠,7♦]}
Dealer stands with 17
Player looses :(
```

CHAPTER 8   BLACKJACK: FULL HASKELL PROGRAM

And so on. The only last comment I'd like to make is let's take one final look at the execStateT function, which has the following type: `Monad m => StateT s m a -> s -> m s`. May seem scary, but since we've long learned how to properly read Haskell type signatures, we can easily decode. In our Blackjack program StateT s m a stands for StateT GameState IO a, s stands for GameState, and m s is then simply IO GameState. This tells us that execStateT expects an action in any monad stack where StateT is on top and an initial state – and then executes this action and returns final state *wrapped in the underlying monad*. This is very important to remember – if you have a stack that has *several* monads, you'll need to unwrap them using corresponding exec (and similar eval and run) methods. Read the docs for StateT, and then you will easily handle Reader and Writer in the same fashion.

IMPROVE BLACKJACK

There is a lot of room for improvement in the current Blackjack implementation. Think of what those can be and implement at least some of them. To get you started:

- Various "don't repeat yourself" optimizations.
- Various checks – for the player's balance and others.
- Add Reader and Writer monads to the stack for better configuration and player notification functionality.
- Add a multiplayer.

The key ideas that we tried to convey by building this simplified Blackjack program from scratch have to do mainly with the approach to designing and building Haskell programs. Let us review them once again – they are absolutely key in making you not just a "coder," but a highly intelligent Haskell developer, who writes code that is easy to read and maintain:

- Design the problem domain using pure types and functions first as much as possible.

- Identify which monads you want to include in the stack of the final program, with State, Reader, Writer, and IO as the usual suspects.

- Build each "floor" of the monad transformer stack separately, using the "the lower the function can live, the better" principle. Remember that lower functions can be lifted, while higher-floor functions are inaccessible from the lower floors.

- As you are building the "floors," expand the pure foundation as well as helper functions from the lower floors to your current one to help your code be more concise and maintainable.

- Connect everything together in the "main" function using a hierarchy of "exec" or "eval" functions from monadic interfaces.

This design pattern, while quite simple, is extremely powerful and can be used to create applications of significant complexity and production quality. With this, we are finishing the first, mainly educational, part of the book and moving to more advanced concepts that we will review with specific practical examples using the principles we have learned so far. Hopefully, it can strengthen the Haskell foundation we built by now and make you fluent in analyzing any new abstractions and approaches you may face on your Haskell journey!

Conclusion

In this chapter, we've embarked on a comprehensive journey to design and implement a simplified Blackjack game using Haskell. This exercise has not only reinforced our understanding of Haskell's functional programming paradigms but also demonstrated the practical application of monads and monad transformers in structuring complex programs.

Key takeaways from this chapter include

- Domain modeling with pure types and functions: We began by modeling our problem domain using pure types and functions. This foundational step ensures that our core logic is both reliable and testable, free from side effects.

- Monad transformer stack design: We explored the design of a monad transformer stack, specifically using StateT over IO, to manage state and side effects. This approach allows us to cleanly separate pure logic from impure operations, maintaining the integrity of our functional design.

- Layered architecture: By organizing our program into distinct layers or "floors," we ensured that each component of our application is responsible for a specific aspect of the program's functionality. This modularity enhances readability, maintainability, and scalability.

- Leveraging Haskell libraries: Throughout the chapter, we utilized Haskell's rich ecosystem of libraries, such as mtl and random, to handle common tasks efficiently. Understanding these libraries and their documentation is crucial for effective Haskell development.

- Iterative development and testing: We emphasized the importance of iterative development, using tools like ghci for interactive testing. This practice helps catch errors early and refine our approach as we build complex systems.

- Opportunities for enhancement: While we implemented a basic version of Blackjack, we identified numerous opportunities for improvement, such as adding more game features, optimizing code, and expanding the monad stack with additional layers like Reader and Writer.

By following these principles and practices, we've not only built a functional Blackjack game but also laid a strong foundation for tackling more complex Haskell projects. As we move forward in the book, we'll delve into advanced concepts and real-world applications, applying the design patterns and techniques we've mastered here.

CHAPTER 9

Let's AI

As I started working on this book, it became impossible to ignore what is happening in the world of AI, or artificial intelligence. Full disclosure: I may be biased as we are currently building a platform for easy creation of AI agents (integrail.ai – and, yes, we do use Haskell for parts of the solution), but still, we may argue about the extent to which AI has influenced and will influence our lives, but we will all agree that it has. In the second part of this book, we will continue building different real-world programs while introducing more and more advanced concepts. Many of these examples will be AI related, on one hand, and, on the other hand, if you use a simple agent or two when learning Haskell and when building programs, you will become *much* more productive than anything that was possible previously. We will build such agents together, in Haskell, for Haskell.

Before we do, we need to make a brief overview of what Agentic AI, enabled by the advances in generative AI and especially the advent of LLMs, or large language models, is.

Agentic AI and Large Language Models

AI is an old term; the topic has always fascinated people. It fell in and out of fashion, depending on the technology available. We had invented "perceptron" – the first ever attempt to create artificial neural networks, or ANNs, over 70 years ago. We had a decade or so of "expert systems" in the 1980s. We witnessed a deep learning revolution that allowed us to train

CHAPTER 9 LET'S AI

multi-layered ANNs in the 2000s. This brought immense advancements in computer vision and enabled "self-driving" vehicles. We had a brief spotlight on so-called Deep Reinforcement Learning with AlphaGo and AlphaZero beating humans at Go and chess – something that seemed impossible for Go not so long ago. Now, it is difficult to find a person who hasn't at least heard of "ChatGPT" and maybe even other "chatbots," and some vendors and individuals dangle the promise of "artificial general intelligence" "sometime soon."

Herein lies a slight problem. Large language models, or LLMs, based on so-called transformers architecture, are an amazing tool. OpenAI with GPT, Anthropic with Claude, Google with Gemini, and Meta with a bunch of excellent and open source Llama models are all based on this same architecture. The good news is using today's level of technology, we can create amazing versatile AI agents that are able to perform a lot of mundane tasks on our behalf, as well as be very helpful in learning and creative endeavors. Bad news: People en masse don't really understand how to utilize LLMs efficiently and are trying to get some overhyped results from the "chatbots" such as ChatGPT, are failing at that, and become disillusioned.

We will try to show a better way in the second part of this book, following the journey of internalizing increasingly more sophisticated Haskell concepts along the way.

We will not be building ANNs or LLMs from scratch in Haskell, even though we could. For better or worse, Python is a de facto standard in AI development, and even though Haskell is very well suited for the architectures of ANNs, that's the reality. So we will focus on *applications* of AI using Haskell, reusing the models that were built in and are being run on Python.

LLMs in a Nutshell

Since this is not a book on ANN architectures, I will mostly focus on practical issues instead of going deep into technical details. In a nutshell, an LLM is an ANN that learns to do one thing only: given the context of a bunch of so-called tokens, it predicts the probability of the next token. That's it. In the last couple of years, there has been very visible progress on improving how LLMs perceive the context first and foremost through so-called "attention" techniques, and this gave us increase in the amount of input information they are able to process while still not being lost. However, the core principle remains the same – predict the probability of the next token based on all the previous.

So how come they can be so efficient at certain tasks, so that people even talk about "reasoning" and the like? If they only learned the information they've seen, we could explain that LLMs are able to answer "tricky" questions such as "how many tennis balls fit into Boeing 747" – since they have seen this question answered in the training data multiple times. But if you ask them "how many children's cubes fit into a Tesla car," they would still give you a very close train of thought, even though I doubt anyone ever formulated this problem this exact way – and this trait gives visibility of their ability to "reason." How is that possible?

The answer lies in another breakthrough in natural language representation studies – so-called vector embeddings. Using some very smart techniques, every word, token,[1] or more complex concepts described as a phrase are converted into multidimensional vectors. It turns out that words and concepts represented this way lie very close in the vector space if they are close in meaning. For instance, a famous example

---

[1] Tokens are several symbols – they can be one short English word or parts of bigger words. We could train LLMs based on whole words – but then they wouldn't learn complex words or nonexistent words so well or based on separate characters – but then there is too much noise. So tokens, such as "un"-"natural," turned out to be an optimal solution.

shows that king - queen = man - woman in the vector space. This means that when LLMs predict probability of the next token and are asked about "children's cubes" instead of "tennis balls" – these concepts lie very close in the vector space, so the "tennis balls" line of reasoning is then applied to the "children's cubes" – we get this impression of the ability to reason (which it is to a certain extent – in a sense, LLM learns "by analogy").

Vector embeddings gave rise to vector databases as well, a very fast developing area in AI, since they allow you to search "by meaning" instead of by keyword. We will discuss the importance of this down the line.

What are LLMs good at?

- Text transformation, following the name of the architecture. Thanks to those vector embeddings, if you want to convert between any formats, they do it very well. Text to JSON or SQL, translation between languages, explaining what software code does in human language, etc. This leads to a number of applications, such as

    - Any integrations.

    - Learning assistance.

    - Translation.

    - Customer support/natural language answering tasks.

    - "Understanding" images. Latest multimodal models do this extremely well, and not surprisingly – after all, it is simply "translation" between pixels and text.

- Text generation, and I mean any kind of text (including code or pretty much anything that can be classified as a language – if they were trained on this data, of course). This one is obvious and is probably the most common use case, used widely in business already.

What are LLM limitations? The two most known arguably are

- So-called hallucinations. When you ask any question of LLMs, they tend to produce an answer no matter what (it's their underlying math after all) and do it very convincingly. So, if we are not an expert in the field, how in the world will we know if what they are telling us is true?!

- Inability to properly model the world knowledge in any "logical" sense of the word. For instance, if you ask LLMs "who is the mother of Tom Cruise," they will tell you, correctly, "Mary Lee Pfeifer." But if you rephrase the question and ask "who is the son of Mary Lee Pfeifer," you won't get an answer. The reason is that they have seen lots of context where Mary Lee was called the mother of Tom Cruise, but very rarely any texts where Tom Cruise was called a son of Mary Lee. And since there is no "internal" logical, reasoning mechanism inside LLMs, this becomes another significant limitation on the way to "true" AI.

As we shall see, the first problem can be solved in a multitude of ways and fairly easily. The second is more difficult, but even with this limitation, LLMs will help us with a number of hard, or at least annoying, problems.

CHAPTER 9 LET'S AI

Three Ways to Make LLMs Better

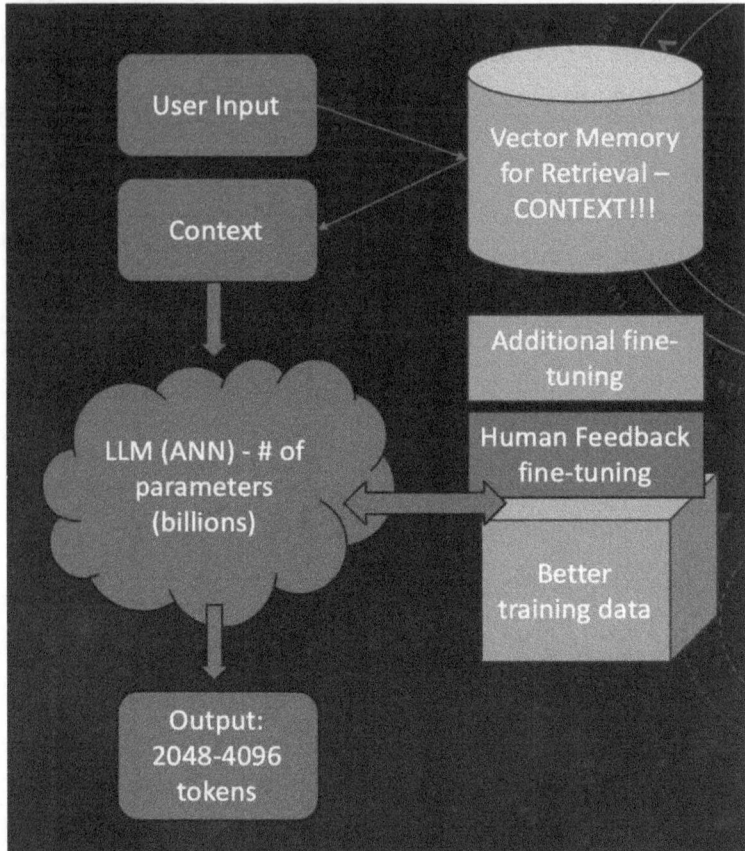

Figure 9-1. Typical LLM and the ways to make it better

Vector RAG

Take a look at Figure 9-1, which shows high-level LLM schematics and the ways to improve their performance. The easiest way to virtually kill hallucinations and provide high-quality output is so-called RAG, or retrieval augmented generation. The way it works is pretty straightforward – whatever the user is asking of the LLM is getting converted into vector embeddings, and a search in the vector database to find pieces of context close in meaning is performed. Then, these relevant pieces of context are being fed as part of the input to the LLM. This way, LLMs stop hallucinating and provide output that follows the meaning in the immediate context. In the worst case, you can always make the LLM respond directly "I don't know" in case it is not 100% sure of the correct answer via prompting. This is the easiest, most straightforward, but very efficient way to make LLMs useful for you or your business personally – build different vectorized knowledge bases, attach them to LLM-based agents, and enjoy much higher quality of the results than with "generic" chatbots from the leading vendors.

Additional Fine-Tuning

All of the available commercial "chatbots" on the market as of this writing have gone through a so-called "human feedback reinforcement learning" stage. It's additional extensive training to make LLMs better suited for a question-and-answer type of conversation, as well as the way to "align" the values and attitudes the LLM learns to what humans want it to have.

However, you can also do additional fine-tuning yourself. Normally, you would need at least several thousands – better yet, tens of thousands – of data samples if you want your fine-tuning to influence the LLM weights. Very unscientifically, I would estimate that proper context building with RAG and other techniques should cover 95% of cases for you, but for some specific cases, where deep vertical knowledge is required, for instance, fine-tuning can be your answer. It also makes sense if you use certain vast context day in and day out – so it will be less expensive to spend once to fine-tune the model instead of wasting tokens at each request.

CHAPTER 9   LET'S AI

Better Original Training Data

Available LLMs are usually trained on everything vendors can get their hands on – Wikipedia, "all of the Internet," books, tweets, GitHub code, what have you. This provides versatility, but also lots of noise. There is research that if you use high-quality training data only in the first place, you need less of it and you need fewer-parameter LLMs to learn it. However, training even a three-billion-parameter model from scratch is a very expensive endeavor, so unless you are a big corporation with very deep pockets, this way is not currently accessible.

Haskell and LLMs

As mentioned in an aside above, we will not be building any neural networks from scratch in Haskell, even though we could. First of all, Python is the language of choice for AI, for better or for worse. Second, even if we do build a simple transformer, e.g., of GPT-2 level, training it properly is very expensive, unpractical, and adds no value in terms of either learning Haskell or doing AI. What we can and will do in Haskell is build a framework with a number of different programs that automate work with different AI agents and that can serve as a solid foundation for you to build your own cool AI agents in Haskell.

We will get there gradually, step by step. First, we will aim to create a "simple" chatbot, powered by various different LLMs, but we will be doing that in frames of a larger-scale design of the whole framework that will allow us to work with our agents via web and via terminal (and operating system in general), as well as having a MongoDB layer to provide persistence for various documents with optional vector search functionality. Then, as we build and flesh out this framework, we will gradually increase the complexity of AI that we will be doing. From the chatbot to multimodal and multi-LLM agents, to integrations with external systems, and to self-learning techniques – unleashing the power of AI agents and your creativity along the way.

Big Picture of the Haskell AI Framework

Here is a high-level description of what we want to have as an end result:

- An "AI agents core":
    - Ability to connect to various AI (generative models) APIs
    - Ability to work with any external APIs to integrate with external systems
    - A lightweight DSL to describe AI agents and build them dynamically at runtime
- Knowledge layer: MongoDB for storage and optional vector search. Why MongoDB and not some SQL database? For one, SQL databases are very well known and well described in the Haskell world as well, so it is more interesting to work with Mongo as a no-SQL leader very widespread in the web programming world. As a bonus, it has very good built-in vector search functionality, which we will need for RAG as described above.
- Different ways to work with the system:
    - Via terminal, using an excellent Haskeline library
    - Via web, with all the consequences of working with JSON, security, etc.
    - Autonomous "agentic mode"

To design and build such a solution, or even something close to it, will require from us not only applying everything we've learned so far but also the knowledge of many other advanced Haskell concepts, such as mutable variables, web servers, multi-threading, vectors and type families, etc. Some of these topics and parts of the framework we will discuss in more detail, and some will only skim merely giving you a direction for further studies or development on your own. But as we progress, I hope that the beauty of Haskell and the typed functional approach to programming in general will captivate you more and more, so that you become a devoted user yourself.

Terminal Chatbot

As they say, before we fly, we gotta crawl. Well, I would argue we learned to crawl in the first part of the book, so let's try at least walking. To do that, we will build a prototype starting point of our AI framework – a simple terminal LLM-powered chatbot. Even though it is simple, it won't be so easy, as we will need to address the following:

- Calling LLMs via API

- Logging errors and general application status (so maybe a Writer monad in the stack?)

- Reading configuration and settings to know how to connect to various APIs (so maybe a Reader monad in the stack?)

- Interacting with the user in the terminal using Haskeline (which provides its' own monad – so integrating it will be a very educational experience)

- Potentially, saving conversation histories and providing additional context

CHAPTER 9　LET'S AI

Let's start!

As we have been discussing and practicing, we need to start thinking about the large-scale program structure (in our case the monad transformer stack, as we don't know any other ways yet), as well as about which functions and types we are going to need and on which floors they should live. As we will use Haskeline as our terminal UI (see Figure 9-2), let's see what kind of interface it provides. Go to hackage.haskell.org, as usual, and search for "haskeline."

Figure 9-2. *InputT transformer, provided by the Haskeline library*

As we can see, InputT is a monad transformer that also has instances for all kinds of IO-related monads, starting with MonadIO itself. This means that all the functionality of IO is available inside InputT. Since

213

we will probably need Reader–Writer–State monads as well, we have a conundrum – where to put InputT in a stack? The consideration is very easy – since it will provide the UI, it has to have access to all of the rest of functionality. As you remember, we can lift functions from the lower floors but not vice versa, so we should put InputT on the top.

Another neat trick provided for us by Haskell's mtl library is this: since Reader–Writer–State are so common a pattern, there is a separate kind of monad called just RWS, which, as you can guess, provides the functionality of all three of these monads. It makes our life much more convenient as well, since dealing with a stack of 5+ monads would require quite a bit of interface-related plumbing to write concise code.

High-Level Design

Thus, preliminarily, we think that the monad stack for our terminal chatbot will look like Figure 9-3.

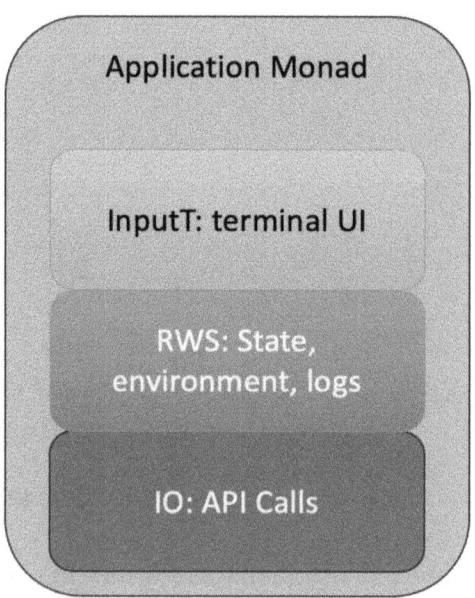

Figure 9-3. *Monad transformer stack for the terminal chatbot*

Now we need to start thinking which functions and types live where. In the first version of our bot, we will use OpenAI API. It is a de facto standard, which other LLM providers try to follow when they define theirs, so it's worth it to review in detail. To use it you will need to get your own OpenAI security token – just navigate to their website, register, and create your own API token. In the next chapters we will discuss alternative ways to use LLMs as well – one possibility is to run new Llama models (Llama-3.2 at the time of this writing) locally, which requires quite a bit of additional Python-related setup, but no API keys.

Here are some other key requirements that we will need to use in our chatbot application:

- Reading configuration and settings from an .env file, using the **dotenv** package

- Converting to/from JSON values, using Data.Aeson from the **aeson** package

- Making HTTP API calls, using the series of functions from the Network.HTTP... namespace, spread out throughout several packages

- Various text and bytestring manipulations

Be ready to have these libraries open on Hackage, along with OpenAI API docs (this is optional as I will explain how to use it in detail).

Initial RWS Monad

Let's start implementing our design gradually. First, we'll create an outline for our RWS Monad by stacking it on top of IO and set up placeholder types for our state and read-only environment variables. We will be gradually extending them as we build out our chatbot.

CHAPTER 9   LET'S AI

You should put the environment variables to the **.env** file in the root folder of the project. You should have a template file called **.env.template** in the repository – copy it to **.env** and fill in the missing info:

```
OPENAI_ORG=
OPENAI_TOKEN=

OPENAI_URL_CHAT=https://api.openai.com/v1/chat/completions
OPENAI_URL_MODELS=https://api.openai.com/v1/models
```

Never commit the .env file to public repositories – as it will contain all the secrets we will use in the project, such as API keys and other authentication-related info.

Our first task is to read the environment and set up our internal, read-only environment in the RWS monad that we will be using. For now, our read-only environment will contain only the list of LLM providers:

```haskell
-- src/StackTypes.hs
...
import LLM.OpenAI (ProviderData)

data Settings = Settings {
    llmProviders :: [ProviderData]
}
```

The ProviderData type will also be built gradually. For now we will use only the following information to set up our initial program:

```haskell
-- src/LLM/OpenAI.hs
...
data ProviderData = ProviderData {
    providerName :: Text,
    providerKey :: BS.ByteString, -- with Bearer for now to
    be faster
    chatCompletionURL :: String
} deriving (Show)
```

We will need the functions to read the environment and configure the initial Settings object, which for now will only contain the settings for openai as a provider:

```haskell
-- src/Init.hs
...
buildOpenAISettings :: IO ProviderData
buildOpenAISettings = do
    -- doing pattern match because we will have checked for
       presence before
    (Just key) <- lookupEnv "OPENAI_TOKEN"
    (Just url) <- lookupEnv "OPENAI_URL_CHAT"
    pure $ ProviderData {
    providerName = "openai",
    providerKey = pack key,
    chatCompletionURL = url,
    providerDefaultOptions = defaultChatOptions
}

initConfig :: IO Settings
initConfig = do
    loadFile defaultConfig
    oa <- buildOpenAISettings
    pure $ Settings { llmProviders = [oa]}
```

There are two straightforward methods: one finds openai-specific configuration in the environment, and the other one reads the environment from the .env file and calls various functions to build the final settings. For now it's only openai. You may also notice that we are taking a shortcut in buildOpenAISettings by pattern-matching directly into (Just x). We are doing so because eventually we will have an additional check that makes sure all needed environment variables are present – and it makes our code here less cumbersome as we don't have to deal with

CHAPTER 9 LET'S AI

Nothing values. The pack function converts the String value that is read from the environment to the ByteString that we will need for our token down the line.

Now let's define our RWS on the IO monad and see how the environment we just read can be used in the main program:

```
-- src/App.hs
...
import Control.Monad.RWS
import StackTypes (Settings, AppState)

type App = RWST Settings [String] AppState IO
```

Here we are defining a monad transformer stack App that consists of the RWS monad with read-only environment type Settings, type for Writer – simple list of Strings (we will change this in the future) – and AppState for state (currently empty), sitting on top of the IO monad.

Now, the challenge beginners often have is this. To execute our program in "main" (even without InputT on top yet), we need to use a function runRWST :: RWST r w s m a -> r -> s -> m (a, s, w) that takes our RWST (or App) action, initial settings, and initial state and then returns IO (a, state, writerOutput). So how do we give the initial settings and state to such a function? Well, you simply run a couple of IO computations that build the initial settings and state for you by reading the disk and network, interacting with the user, etc. and then pass them to runRWST. Again, if you carefully read and decode the types, it becomes very easy to understand. For instance, for our very simple current program state, we could write main something like the following:

```
main :: IO ()
main = do
    settings <- initConfig
    print settings
    runRWST mainAction settings initialState
```

The logic is you build the read-only settings once before the program is initiated, and then they become read-only and pure for all practical purposes inside whatever crazy monad transformer stack you may have.

The above illustrates how you can start with a pretty general application skeleton, including your monad stack, and then gradually build it by expanding the state and environment types, as well as designing the problem domain types and functions as usual. Let us follow our own advice and move to the following.

Designing the OpenAI Interface

In this section, we will dive into the details of how we interface with OpenAI's APIs to build a seamless connection for our chatbot. The goal is to abstract out the complexities of the API, allowing Haskell users to easily send messages, receive responses, and manage configuration options, all while keeping everything typed and functional.

We will cover a few key aspects that are central to working with OpenAI's API using Haskell:

- **Converting to and from JSON** for the API request/response
- **Handling OpenAI message formats**
- **Sending requests to the API** for chat completion

Converting Between Haskell Types and JSON

In order to interact with OpenAI's API, our Haskell types need to be converted to and from JSON. OpenAI requires requests to be sent in a specific JSON format, and responses are returned in JSON as well. Thankfully, Haskell's aeson library provides us with powerful tools to make this conversion seamless by leveraging typeclasses like ToJSON and FromJSON.

The Message Type

The Message type represents a single message in the chat interaction. When working with OpenAI models like GPT, you typically send a list of messages with different roles (user, assistant, or system). Here's our Message type:

```
data Message = Message
  { role :: Text -- "user" | "assistant" | "system"
  , content :: Text
  , tool_calls :: Maybe [Text]
  , tool_call_id :: Maybe Text
  } deriving (Show, Generic)

instance ToJSON Message
instance FromJSON Message
```

The role field determines who is sending the message (the user, the assistant, or an internal system message). The content holds the actual text of the message. Optionally, you can include tool_calls and tool_call_id for more advanced interactions, such as invoking external tools during conversations, but we won't be using those for now.

By deriving Generic, we can automatically implement the ToJSON and FromJSON instances, allowing the aeson library to handle the JSON conversion for us.

ChatOptions: Configuring the Chat Interaction

Chat interactions with OpenAI models can be fine-tuned through various configuration options. These options are sent as part of the API request and allow you to control behavior such as the length of the response, temperature (how predictable the model output should be), whether the response is streamed, and more:

```
data ChatOptions = ChatOptions {
    stream :: Bool,
    frequency_penalty :: Maybe Float,
    logprobs :: Maybe Bool,
    top_logprobs :: Maybe Int,
    max_tokens :: Maybe Int,
    presence_penalty :: Maybe Float,
    response_format :: Maybe Object,
    seed :: Maybe Int,
    stop :: Maybe [Text],
    stream_options :: Maybe Object,
    temperature :: Maybe Float,
    top_p :: Maybe Float,
    tools :: Maybe [Object],
    tool_choice :: Maybe Text,
    user :: Maybe Text
} deriving (Show, Generic)

instance ToJSON ChatOptions
instance FromJSON ChatOptions
```

This ChatOptions type models the configurable settings that the OpenAI API accepts. For example:

- stream allows response streaming, where chunks of the response are delivered as soon as they're ready.

- temperature controls the randomness of the model's responses. Higher values (e.g., 0.8) result in more random responses, while lower values (e.g., 0.2) make the model more focused and deterministic.

CHAPTER 9   LET'S AI

We also provide a defaultChatOptions to simplify using the API without the need to manually specify all the options from scratch. This default configuration is geared toward enabling streaming responses, with a temperature of 1 (balanced creativity) and no specific penalties or stopping conditions.

Working with JSON is ubiquitous when developing any web-related applications, so let's spend some time discussing the aeson library. It provides a set of tools to easily convert between Haskell types and their JSON equivalents, as well as to parse and generate JSON efficiently.

The core of aeson revolves around two typeclasses:

- ToJSON: Defines how a Haskell type can be serialized into JSON

- FromJSON: Defines how JSON can be deserialized into a Haskell type

When you have a Haskell value that you want to send as JSON (e.g., in an API request), you need to convert it into a Value (another type provided by aeson that represents JSON). Similarly, when you receive JSON (e.g., an API response), you need to convert it back into your Haskell types using FromJSON.

How ToJSON and FromJSON Work

The easiest way to leverage aeson for JSON conversion is through the **automatic derivation** of ToJSON and FromJSON instances. Haskell's GHC. Generics allows us to automatically derive these instances for most simple data types, provided that the types involved are also instances of ToJSON and FromJSON.

Let's revisit our Message type:

```haskell
{-# LANGUAGE DeriveGeneric #-}
import GHC.Generics (Generic)
import Data.Aeson (ToJSON, FromJSON)

data Message = Message
  { role :: Text   -- "user", "assistant", "system"
  , content :: Text
  , tool_calls :: Maybe [Text]
  , tool_call_id :: Maybe Text
  } deriving (Show, Generic)

instance ToJSON Message
instance FromJSON Message
```

Here's what's happening:

1. We derive the **Generic** typeclass for Message. This allows Haskell's compiler to automatically generate a "generic representation" of the type. Unfortunately, we cannot go into the details of this mechanism here; suffice it to know that it allows Haskell to then do lots of automatic manipulations with your types that saves boilerplate.

2. By deriving **Generic**, we can automatically derive ToJSON and FromJSON instances using aeson. This means we don't need to manually write code to convert a Message to/from JSON – huge timesaver!

With this in place, we can now **serialize** (ToJSON) and **deserialize** (FromJSON) Message values without any additional effort.

Here's how you would convert a Message into a JSON Value:

```
import Data.Aeson (encode)
```

```
let message = Message "user" "Hello, OpenAI!" Nothing Nothing
let jsonMessage = encode message   -- JSON representation of
the message
```

On the flip side, deserializing JSON back into a Message looks like this:

```
import Data.Aeson (decode)
import Data.ByteString.Lazy.Char8 (pack, unpack)
```

```
let jsonString = "{\"role\": \"user\", \"content\": \"Hello, OpenAI!\"}"
let maybeMessage = decode (pack jsonString) :: Maybe Message_
```

If the JSON string matches the structure of Message, decode will return Just Message; otherwise, it will return Nothing.

By default, aeson will map fields in your data structure to JSON key-value pairs based on the **field names**. For example

```
Message { role = "user", content = "Hi", tool_calls = Nothing,
tool_call_id = Nothing }
```

will be encoded into

```
{
  "role": "user",
  "content": "Hi",
  "tool_calls": null,
  "tool_call_id": null
}
```

This is great for most use cases, but what if you have slightly different requirements? For instance, what if the API expects snake_case keys instead of camelCase or you want to rename some of the fields?

In such cases, you can customize how the conversion works by providing your own implementations of ToJSON and/or FromJSON.

Let's say we want the Message type to be serialized with snake_case field names instead of the default camelCase:

```
{-# LANGUAGE OverloadedStrings #-}

import Data.Aeson (ToJSON, FromJSON, object, (.=),
withObject, (.:))

instance ToJSON Message where
  toJSON (Message role content toolCalls toolCallId) =
    object [ "role" .= role
           , "content" .= content
           , "tool_calls" .= toolCalls
           , "tool_call_id" .= toolCallId
           ]

instance FromJSON Message where
  parseJSON = withObject "Message" $ \v -> Message
    <$> v .: "role"
    <*> v .: "content"
    <*> v .: "tool_calls"
    <*> v .: "tool_call_id"
```

Here, we're manually specifying the JSON object structure in both directions:

- **toJSON** converts a Message into a JSON object with the desired field names.
- **parseJSON** converts a JSON object into a Message by extracting values from the corresponding keys.

This approach is useful when you need fine-grained control over the JSON structure, such as when integrating with APIs that have specific field naming conventions.

Handling Optional Fields

In the Message type, some fields are optional (e.g., tool_calls and tool_call_id, which are Maybe types). aeson handles this gracefully by encoding Nothing as null in JSON and decoding null back into Nothing:

```
let message = Message "user" "Hello!" Nothing Nothing
let jsonMessage = encode message
```
This will produce:
```
{
  "role": "user",
  "content": "Hello!",
  "tool_calls": null,
  "tool_call_id": null
}
```

When aeson encounters null in JSON, it correctly maps it back to Nothing in Haskell during decoding.

aeson in Action: API Requests and Responses

Now that we understand how ToJSON and FromJSON work, let's see how they fit into the broader context of interacting with OpenAI's API.

In the **chatCompletion** function, we send a POST request to OpenAI's chat completion API. To do so, we first need to prepare the request body, which consists of the model ID, the list of messages, and the optional chat options.

Using the ToJSON instances, we can easily convert these Haskell values into JSON for the request body:

```
let obj = Data.Aeson.object [
        "model" .= modelId,
        "messages" .= toJSON messages
    ]
let request = setRequestBodyJSON (combineObjects obj
(toJSON fopts))
            $ applyBearerAuth key initialRequest
```

Here, we create a JSON object using the **object** function provided by aeson, which takes a list of key-value pairs. The .= operator is used to associate a key with its value (converted to JSON via toJSON). This allows us to easily build the JSON structure required for the API request.

When we receive a response from the API, it will be in JSON format. Using the **FromJSON** instance for OpenAIResponse, we can automatically decode the response body back into a Haskell value:

```
let jsonObject = decode (BL.pack (unpack ch)) :: Maybe
OpenAIResponse
```

If the response JSON matches the structure of OpenAIResponse, decode will return Just OpenAIResponse; otherwise, it will return Nothing. This makes decoding responses safe and efficient, as we can handle the failure case when the JSON doesn't match.

The OpenAIResponse type models the response from OpenAI's API:

```
data OpenAIResponse = OpenAIResponse
  { id :: Text
  , object :: Text
  , created :: Int
  , model :: Text
  , choices :: [Choice]
  , usage :: Maybe Usage
  } deriving (Show, Generic)
```

CHAPTER 9   LET'S AI

```
instance ToJSON OpenAIResponse
instance FromJSON OpenAIResponse
```

Again, by deriving Generic, we can easily generate ToJSON and FromJSON instances. This allows us to seamlessly convert the entire response into a Haskell structure that we can work with in our application.

Conclusion on aeson

The **aeson** library provides a powerful and flexible mechanism for working with JSON in Haskell. By deriving ToJSON and FromJSON, we can almost always rely on automatic, type-safe conversions between Haskell data types and JSON. When needed, we can customize the serialization/deserialization process to fit specific requirements, such as renaming fields or handling optional values.

For our OpenAI interface, aeson allows us to effortlessly convert Haskell data (such as messages and options) into valid JSON for API requests and decode the responses back into Haskell types. This keeps our code clean, type-safe, and easy to modify as our API interactions evolve.

I encourage you to read the library documentation and experiment with different APIs to get a hang of it. Hopefully, the description above provides enough for you to start.

Sending API Requests: The chatCompletion Function

The core of our OpenAI interaction revolves around sending messages to the model and processing its response. For this, we define the chatCompletion function, which builds and sends the request to OpenAI's API.

In the chatCompletion function, we use the http-client and http-conduit libraries to build a POST request and send it to OpenAI's chat completion endpoint. The request includes

CHAPTER 9 LET'S AI

- The **model ID** (e.g., "gpt-4o") determining which model to use
- The **message** list, which contains the conversation history
- The **chat options**, which control response parameters like streaming, temperature, and maximum tokens

Here's the full function:

```
chatCompletion :: Manager -> [Message] -> Text -> ProviderData ->
Maybe ChatOptions -> (BS.ByteString -> IO()) -> IO ()
chatCompletion mgr messages modelId provider opts func = do
    let url = chatCompletionURL provider
    let key = providerKey provider
    let fopts = Data.Maybe.fromMaybe (providerDefaultOptions
    provider) opts
    initialRequest <- parseRequest url
    let obj = Data.Aeson.object [
                "model" .= modelId,
                "messages" .= toJSON messages
            ]
    let request = setRequestBodyJSON (combineObjects obj
    (toJSON fopts))
            $ applyBearerAuth key initialRequest

    let postRequest = request { method = "POST" }

    httpSink postRequest $ \response -> do
        if statusCode (responseStatus response) /= 200
            then do
                liftIO $ print (responseStatus response)
                liftIO $ print (responseHeaders response)
            else mapM_C func
```

Let's break this down:

- Building the URL and authentication: The chatCompletionURL and providerKey are part of the ProviderData that contains the necessary information to interact with OpenAI. These include the endpoint URL and the authorization token (which we send via the applyBearerAuth function).

- Building the JSON request: The request is built using a **JSON object** containing the modelId and the messages. We combine this message object with the ChatOptions using combineObjects, ensuring both are included in the request body. We discussed this in detail in the previous section.

- Sending the request: Using httpSink, we perform the actual POST request. The response is handled in chunks, allowing us to process streaming results. This is useful when stream is set to True, as the model's response is streamed in real time.

- Processing the response: If the API returns a status code other than 200, we log the error (both the status code and the response headers). Otherwise, we pass each chunk of the response to the provided func callback for further processing.

Handling the Streaming Response

When the model responds in a streaming fashion (controlled by the stream option), the response is delivered as chunks of ByteString. We can process these chunks as they arrive by providing a function that handles each chunk, for example:

```
testChunk :: BS.ByteString -> IO ()
testChunk ch = do
    putStrLn $ unpack ch
    let jsonObject = decode (BL.pack (unpack ch)) :: Maybe
    OpenAIResponse
    print jsonObject
```

This function decodes each chunk as an OpenAIResponse, allowing us to inspect the intermediate output before the full response is complete. The topic of streams and their handling is a vast and deep area that is well covered by other books and tutorials, so we will not go in detail on how to use the conduit library to work with streams – but I do encourage you to study it similar to aeson. With these two libraries in your "spell book," you will be able to write high-performance, robust web-related code for any task. Here, we provide only a basic starting point.

Conclusion

We laid a great foundation for our AI chat agent in this chapter by doing initial rough design of the monad transformer stack for the app and writing some code for interacting with OpenAI and the environment. In this chapter's code, we can test that everything works by executing the following main program (app/Main.hs):

```
{-# LANGUAGE OverloadedStrings #-}

module Main (main) where

import Init (initConfig)
import StackTypes (findProviderByName)

import Network.HTTP.Client (Manager, newManager)
import Network.HTTP.Client.TLS (tlsManagerSettings)
```

CHAPTER 9   LET'S AI

```
import LLM.OpenAI (chatCompletion, userMessage,
testChunkStreaming)

main :: IO ()
main = do
    settings <- initConfig
    let prov = findProviderByName settings "openai"
    print prov
    manager <- newManager tlsManagerSettings
    chatCompletion manager [userMessage "hi"] "gpt-4o" prov
    Nothing testChunkStreaming
```

For now, we do not use our monad transformer stack at all – we will build it out in the next chapter. We just want to check in the good old IO that our OpenAI API interaction code is working. If you set up the OpenAI token correctly, you should see something like the following after running `stack build`, `stack exec bot`:

```
{19:04}~/dev/magicalhaskell/ch09:main X ↪ stack exec bot
ProviderData {providerName = "openai", providerKey = "sk-
proj-G4g2bmbFcIkOR1EEdiU2T3BlbkFJvmmBpQwSYxdhyOnf6V
hg", chatCompletionURL = "https://api.openai.com/v1/chat/
completions", providerDefaultOptions = ChatOptions {stream =
True, frequency_penalty = Nothing, logprobs = Nothing, top_
logprobs = Nothing, max_tokens = Nothing, presence_penalty =
Nothing, response_format = Nothing, seed = Nothing, stop =
Nothing, stream_options = Nothing, temperature = Just 1.0,
top_p = Nothing, tools = Nothing, tool_choice = Nothing, user =
Nothing}}

Length is: 0
Nothing
```

CHAPTER 9 LET'S AI

```
{"id":"chatcmpl-ADE6iQUdKNFMlJmmNmhOB6M9ZJLex","object":
"chat.completion.chunk","created":1727715856,"model":"g
pt-4o-2024-05-13","system_fingerprint":"fp_057232b607","choices
":[{"index":0,"delta":{"role":"assistant","content":"","refusal
":null},"logprobs":null,"finish_reason":null}]}
```

Length is: 286
Just (OpenAIResponse {id = "chatcmpl-ADE6iQUdKNFMlJmmNmhOB6M9ZJLex", object = "chat.completion.chunk", created = 1727715856, model = "gpt-4o-2024-05-13", choices = [Choice {message = Nothing, delta = Just (DeltaMessage {content = "", tool_calls = Nothing}), index = Just 0, finish_reason = Nothing, logprobs = Nothing}], usage = Nothing})

```
{"id":"chatcmpl-ADE6iQUdKNFMlJmmNmhOB6M9ZJLex","object":
"chat.completion.chunk","created":1727715856,"model":"g
pt-4o-2024-05-13","system_fingerprint":"fp_057232b607","choices
":[{"index":0,"delta":{"content":"Hello"},"logprobs":null,"fini
sh_reason":null}]}
```

Length is: 257
Just (OpenAIResponse {id = "chatcmpl-ADE6iQUdKNFMlJmmNmhOB6M9ZJLex", object = "chat.completion.chunk", created = 1727715856, model = "gpt-4o-2024-05-13", choices = [Choice {message = Nothing, delta = Just (DeltaMessage {content = "Hello", tool_calls = Nothing}), index = Just 0, finish_reason = Nothing, logprobs = Nothing}], usage = Nothing})
...

...proving that everything works and we can indeed listen to the streaming response coming back from OpenAI API.

CHAPTER 9   LET'S AI

Conclusion

In the next chapter, we will build up our monad transformer stack gradually to make our chat agent nice, useful, scalable, and a good foundation for your further experiments. We will also build a nicer OpenAI API so that we don't need to deal with separate "chunks." Then we will also expand the list of possible LLMs and agent functionality to work with and eventually will expand our framework with the web access functionality (in Chapter 11).

> **IMPROVING THE CHAT AGENT EXERCISES**
>
> 1. Improve the OpenAI stream management function so that it creates a nice response message instead of logging all the meta information.
>
> 2. Start building the "State" floor in our monad transformer stack to keep track of the current conversation with the chat agent – keeping message history would be a good starting point.

In this chapter, we've embarked on an exciting journey into the world of AI, specifically focusing on leveraging large language models (LLMs) to build a terminal-based chatbot using Haskell. This endeavor not only introduces us to the practical applications of AI but also deepens our understanding of Haskell's advanced concepts and libraries.

Key takeaways from this chapter include

- Understanding LLMs and their applications: We explored the capabilities and limitations of LLMs, such as text transformation, generation, and their potential for reasoning through vector embeddings. This foundational knowledge helps us appreciate the power and constraints of AI technologies.

- Designing a monad transformer stack: We laid the groundwork for our chatbot by designing a monad transformer stack that includes RWS on top of IO, with InputT from the Haskeline library at the top. This design allows us to manage state, configuration, and user interaction effectively.

- Interfacing with OpenAI's API: We delved into the practical aspects of interacting with OpenAI's API, including setting up authentication, sending requests, and handling streaming responses. This involved using Haskell's aeson library for JSON conversion and http-client for HTTP requests.

- Leveraging Haskell's type system: Throughout the chapter, we utilized Haskell's strong type system to ensure safe and efficient conversions between Haskell types and JSON, demonstrating the power of type safety in real-world applications.

- Incremental development: We adopted an incremental approach to building our chatbot, starting with basic functionality and gradually expanding it. This method allows us to test and refine each component before moving on to more complex features.

In the next chapter, we'll further develop our chatbot, making it more robust and scalable, and lay the foundation for integrating web access and additional AI functionalities. This will set the stage for creating a comprehensive AI framework that can serve as a powerful tool for both learning and practical applications.

CHAPTER 10

Terminal AI Chat Agent

In this chapter, we will finish our AI chat agent – let's be unoriginal and call it "Jarvis" – to a full-featured terminal-based application. This will give us an opportunity to apply everything we've learned so far on the next level while creating something cool and exciting. Figure 10-1 shows a monad transformer stack we decided to use for our application.

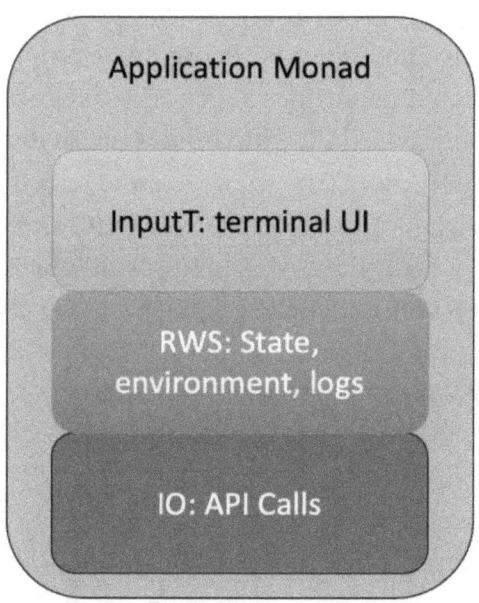

Figure 10-1. *Monad transformer stack for Jarvis*

CHAPTER 10   TERMINAL AI CHAT AGENT

In the previous chapter we started designing the types and functions that allow us to work with the OpenAI API. They were all either pure or in the IO monad. We also defined a part of our monad stack using the RWST monad transformer – a cool abstraction, which combines Reader, Writer, and State monads into one. In this chapter, we will design and build all the missing pieces to make Jarvis functional.

The code for this chapter is in the "ch10" folder of the book's GitHub repository.

Terminal UI Skeleton

Let us start with finishing the top-level program structure by adding the "top floor" of our transformer stack – InputT monad transformer from the Haskeline package. As always when we use external libraries, it's a good idea to go to hackage.haskell.org, search for it, and read the docs as needed. Haskeline provides a nice and easy-to-use interface to working with terminal-based UI, with things such as history tracking and autocomplete out of the box. To start, we want to have a fully functional program that takes user input, goes to OpenAI API, and prints a response. We can create it very easily now, designing which functions and types should live on which floors of our monad stack, as we did in the previous chapters. To do this, we'll need to restructure our files a little bit, further updating our architecture illustration (Figure 10-2).

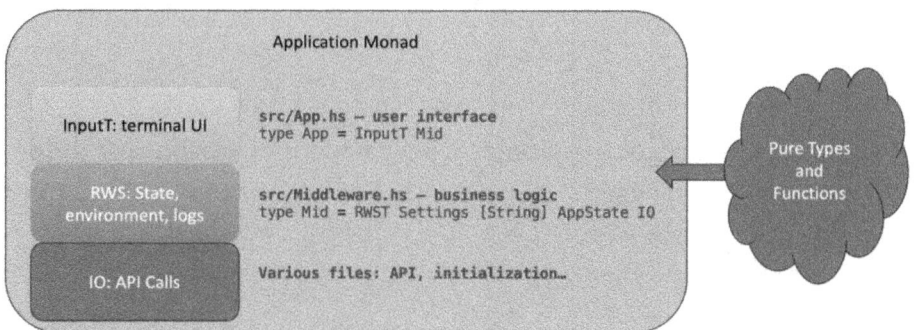

Figure 10-2. *Current Jarvis architecture*

As we shall soon see in detail, we again are applying the design approach from the previous chapters, where we clearly separate types and functions in separate "worlds": pure for everything that we *can* make pure, IO "floor" for all kinds of functions that help us interact with the external world (such as OpenAI API), and RWS floor we now moved to the Middleware.hs file (used to be App.hs in the last chapter) – and it will contain all functions that deal with the business logic of the application and that require keeping state and reading our settings, but *do not* require direct interaction with the user. The UI actions will live on the top floor, inside the InputT monad in the src/App.hs file. Let us review the functions that are required to make a bare minimum of the interactive AI chatbot, in addition to the foundation we laid in the previous chapter.

Adding a Nicer Response Function

First, let's change the function that processes the streaming OpenAI API response so that it gives us a nice textual message instead of all the JSON meta information as in the last chapter:

```
-- src/LLM/OpenAI.hs
...
processResp :: BS.ByteString -> IO ()
```

```
processResp ch = do
  mapM_
    ( \x -> do
        let jsonObject = decode (BL.pack x) :: Maybe
        OpenAIResponse
        case jsonObject of
          Nothing -> return ()
          Just resp -> do
            let chs = delta $ head $ choices resp
            case chs of
              Nothing -> return ()
              Just DeltaMessage {content = cnt} -> do
                putStr (T.unpack cnt)
    )
    (processString $ unpack ch)
```

This is the function that we will pass to the call to OpenAI API streaming response instead of the `testChunkStreaming` we used in the last chapter. What we do here is split the stream string with `processString` to get rid of the "data:" prefix to get to the list of JSON objects we are looking for. Various calls to `pack` and `unpack` have to do with conversions between different string formats – Text, ByteString, String, etc. This may get a little tedious to write, but with accurate reading of the type signatures, you'll get a hang of it fast. Then, we `mapM_` a "lambda function" defined right in the code over the resulting list of strings. `mapM_` is simply a variant of mapping over a list, which allows us to map IO actions (instead of pure functions) while discarding the results – here we are only interested in side effects of putting characters on the screen. The mapped function itself converts JSON into our internal OpenAI response representation and then finds the needed textual response and outputs it directly to the terminal. Here we are cheating a bit by using an IO action (putStr) to put the characters on the screen – in the finished app we will have to make it a

App() action so that we keep track of the conversation history in the state and use proper Haskeline functions for input/output, but now we just want to get to the result as fast as possible.

> **OPTIMIZE THE PROCESSRESP FUNCTION**
>
> The processResp function now has several case statements because we are dealing with Maybe values. Optimize its code to get rid of the case statements, for instance, by using the maybe function or fromMaybe operator.

Initialization

We need a bunch of initialization functions that read the environment and create the initial Settings object (that is used in the Read part of our RWS monad) and initial state AppState object (that is used in the State part of our RWS monad). To do that, we use IO (this is important). We cannot initialize those objects from inside the actual RWS monad – they are used as initial values in the call to our overall program – and we cannot do it "purely" either, since we need to read the environment. We defined the functions that initialize the settings in the src/Init.hs in the previous chapter. Now let's extend it to also initialize the state:

```
-- src/StackTypes.hs
...
-- type that will hold the State for our App
data AppState = AppState {
    httpManager :: Manager,
    currentModelId :: Text,
    currentProvider :: ProviderData
}
```

```
-- src/Init.hs
...
-- main initialization function in IO to create initial State
and Settings
initAll :: IO (Settings, AppState)
initAll = do
    settings <- initConfig
    let prov = findProviderByName settings "openai"
    print prov
    manager <- newManager tlsManagerSettings
    let initSt = AppState manager "gpt-4o" prov
    pure (settings, initSt)
```

The initAll function currently calls initConfig to create our initial Settings object by reading the environment; then it builds the initial state by putting a default LLM provider (currently we have only one, OpenAI) and an HTTP manager object to help us make external API calls into the initial AppState. Then we "return" both as a pair.

Building the Monad Transformer Stack

The two top floor of our monad stack now can be constructed as well. The middleware, or Mid monad, doesn't have any actions so far, but we need to at least create the monad itself. So the src/Middleware.hs file is very simple now:

```
{-# LANGUAGE OverloadedStrings, DuplicateRecordFields #-}

module Middleware
where

import Control.Monad.RWS
```

CHAPTER 10   TERMINAL AI CHAT AGENT

```
import StackTypes (Settings, AppState (AppState),
 findProviderByName)

-- monad that handles all application's business logic
type Mid = RWST Settings [String] AppState IO
```

We simply define Mid as an RWST monad transformer on top of IO that uses Settings for the Read monad, AppState for the State monad, and [String] – list of strings – for the Writer monad. The latter will be changed down the line, but we need to give some Monoid to the writer so starting with the simple one.

This definition alone, even though it's just one line, allows us to use all the functions of Reader, Writer, and State monads.

Let's move to the top, UI floor defined in the src/App.hs file:

```
{-# LANGUAGE OverloadedStrings, DuplicateRecordFields #-}
{-# LANGUAGE DeriveGeneric #-}

module App
where

import qualified System.Console.Haskeline as HL
import qualified Util.PrettyPrinting as TC
import Control.Monad.RWS
import Network.HTTP.Client (Manager)
import qualified Data.Text as T
import LLM.OpenAI

import Middleware (Mid)
import StackTypes

type App = HL.InputT Mid

loop :: App ()
loop = do
```

```
st <- lift get
let mgr = httpManager st
let modelId = currentModelId st
let pr = currentProvider st
minput <- HL.getInputLine  (TC.as [TC.bold] "\nλInt. ")
case minput of
  Nothing -> return ()
  Just ":quit" -> return ()
  -- Just (':':cmds) -> (processCommand (words cmds)) >> loop
  Just input -> do
    -- outputStrLn $ "Input was: " ++ input
    liftIO $ chatCompletion mgr [userMessage $ T.pack input]
    modelId pr Nothing processResp
    loop
```

For now, it contains one key function that is an action in the App monad – loop. This is a main loop of our application, which we will be gradually extending. For now, it does the following:

- st <- lift get: We are getting the state from the underlying RWS monad. Note the use of "lift" – since we are currently on the "top floor" of our monad stack and the State functionality is provided by RWS, we need to lift the actions from that floor into ours!

- In the next three lines, we bind various items that we will need to call the OpenAI API to specific variables.

- minput <- ...: Here we both output the prompt for our program and expect the input from the user.

- The rest of the code deals with analyzing the user input. Now we only have two options:

  - :quit is a command to quit from Jarvis.

  - Any other input is treated as a message to the AI, which is done through another lifted call liftIO $ chatCompletion ... - we looked at the chatCompletion function in detail in the last chapter, so feel free to review that discussion if you need to.

Main Program

We have all the pieces we need to make an actual program. All that is left is to construct our main function. It might seem tricky at first, as with all monad transformer stacks, but with time it will become your second nature:

```
{-# LANGUAGE OverloadedStrings #-}

module Main (main) where

import Init (initAll)
import StackTypes (findProviderByName)

import Network.HTTP.Client (Manager, newManager)
import Network.HTTP.Client.TLS (tlsManagerSettings)
import LLM.OpenAI (chatCompletion, userMessage,
  testChunkStreaming, processResp)
import System.Console.Haskeline
import App (loop)
import Control.Monad.RWS
```

CHAPTER 10   TERMINAL AI CHAT AGENT

```
main :: IO ()
main = do
    (sett, initSt) <- initAll
    runRWST (runInputT (defaultSettings {historyFile = Just
    ".jarvis_history"}) loop) sett initSt
    putStrLn "bye!"
```

Apart from all the imports, the main function is actually very, very simple. We initialize the initial Settings and AppState objects using initAll – an IO action we discussed above. Then we run our monad stack. Let's unwind this line so that you understand how easy it is despite looking intimidating. To do that, we need the signatures for runRWST and runInputT:

```
runInputT :: (MonadIO m, MonadMask m) => Settings m -> InputT
m a -> m a
runRWST :: RWST r w s m a -> r -> s -> m (a, s, w)
```

Now you can read the line from our main function as follows: `runRWST <some action> sett initSt`. Here sett is our Settings object, initSt our initial state, and the result will be `m(a,s,w)` or in our case `IO ( (), AppState, [String])` – an IO action. As we are inside main, which is in the IO monad, everything is fine. Now let's deal with <some action>. runRWST expects `RWST r w s m a` type for it – looks scary, but let's make the substitutions and we will see it is merely `RWST Settings [String] AppState IO ()` type, but if you recall our definition of Mid, it's exactly `Mid ()`! So <some action> has to have this type. Now let's see what it is `runInputT (defaultSettings {historyFile = Just ".jarvis_history"}) loop`. runInputT expects some Settings object, and here we are giving a defaultSettings object provided by Haskeline only with the history file name changed and then `InputT m a` action. If you recall, our

CHAPTER 10   TERMINAL AI CHAT AGENT

App is defined as InputT Mid, and loop has the type App(), i.e., InputT Mid ()! That's it. Everything typechecks. You just have to carefully go through the exercise of unwinding monad transformer stacks a couple of times, and then you will be able to do it in your sleep.

Now if you run the whole program with the changes above via stack build, stack exec jarvis, you should see something like

```
{11:25}~/dev/magicalhaskell/ch10:main X ⇨ stack exec jarvis
```

ProviderData {providerName = "openai", providerKey = "…", chatCompletionURL = "https://api.openai.com/v1/chat/completions", providerDefaultOptions = ChatOptions {stream = True, frequency_penalty = Nothing, logprobs = Nothing, top_logprobs = Nothing, max_tokens = Nothing, presence_penalty = Nothing, response_format = Nothing, seed = Nothing, stop = Nothing, stream_options = Nothing, temperature = Just 1.0, top_p = Nothing, tools = Nothing, tool_choice = Nothing, user = Nothing}}

λ**Int.** hi
Hello! How can I assist you today?
λ**Int.** is Haskell a good language?
Whether Haskell is a "good" programming language largely depends on your specific needs, goals, and preferences. Haskell is known for several distinctive features that can make it highly attractive in certain contexts:

Pros of Haskell:

...

Pretty cool, eh? We already have a functional chatbot, even though a very limited one – it does not track the message history, we took some shortcuts when defining API functions, etc. However, what we do have is

CHAPTER 10   TERMINAL AI CHAT AGENT

a full monad transformer stack, core API functions, and a fully functional program. Now we can build up some muscles around this skeleton and create a program of pretty much any complexity.

MESSAGE HISTORY EXERCISE

Think how you could approach tracking the message history in our chat agent. For this, we would need to change the processResponse function as well as amend our AppState to store everything properly before calling OpenAI API.

Building Muscles

So what else do we want in our program? In this chapter, let's focus on adding three interesting features:

- Keeping track of the message history, so that our conversations have context
- Logging to a separate log file
- Calculating usage to utilize the Writer monad functionality

Let's start with logging – as it will be used by all the other features. We need to define our logging functions first to avoid having to adjust our core API twice.

Logging

We could of course write some kind of basic logger from scratch, but there are lots of very good libraries out there – and logging is not the type of functionality where we want to reinvent the wheel, so let's just use what's available. We will use the **fast-logger** package. As usual, go to hackage.

CHAPTER 10   TERMINAL AI CHAT AGENT

haskell.org and search for it to read the docs. This is an extremely fast logger that also scales well on multiple cores – which is actually one of the very clear strengths of Haskell in general, but outside the scope of this book. However, since we are building a good foundation for potential production solutions, it makes sense to use it from the start – even though we won't utilize its power to the full extent.

The code for the logger (and supporting terminal formatting functions) is in the folder src/Util. We are using fast-logger for the logging functionality, and on top of it we define the usual log levels, convenience functions to log in various levels – such as lg_err, lg_dbg, etc. – and additional state for housekeeping. Then there are a couple of initialization functions, which we will need to add to our initialization routine in main.

You can test how the logger outputs different level messages by loading the code into ghci by running stack ghci and then doing ghci> initLogger VERBOSE >>= testLogs. You should see output similar to that in Figure 10-3.

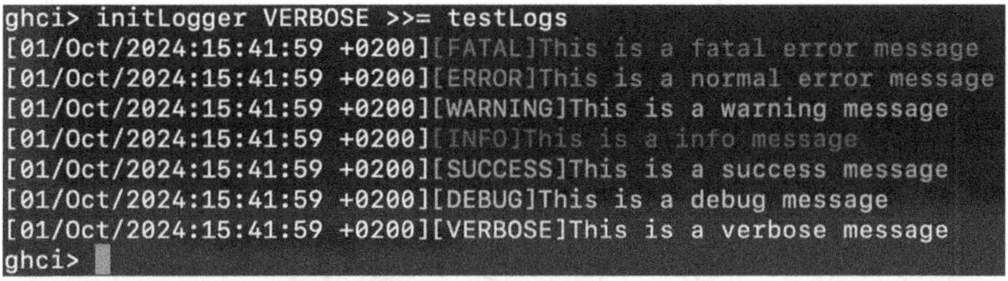

```
ghci> initLogger VERBOSE >>= testLogs
[01/Oct/2024:15:41:59 +0200][FATAL]This is a fatal error message
[01/Oct/2024:15:41:59 +0200][ERROR]This is a normal error message
[01/Oct/2024:15:41:59 +0200][WARNING]This is a warning message
[01/Oct/2024:15:41:59 +0200][INFO]This is a info message
[01/Oct/2024:15:41:59 +0200][SUCCESS]This is a success message
[01/Oct/2024:15:41:59 +0200][DEBUG]This is a debug message
[01/Oct/2024:15:41:59 +0200][VERBOSE]This is a verbose message
ghci>
```

Figure 10-3. *Testing our logger*

If you read the src/Util/Logger.hs file, you will notice that our functions there are all in the IO, and so require an explicit state object, since the type signatures all look like this:

lg :: LogLevels -> [Char] -> [String] -> LoggerState -> IO ()

CHAPTER 10   TERMINAL AI CHAT AGENT

This is of course not what we want to write every time, so we need to create proper logging actions inside the Mid monad – and if we want to use logging in the top floor, or the App monad, we can always lift them.

To do that, first we need to extend the AppState type to keep track of the logger state:

```
-- src/Util/Logger.hs
...
-- logger-specific state, keeping the current logger (can log
to stdout or files, depending on which initialization function
we use, and the clean up function

data LoggerState = LoggerState {
    logger::TimedFastLogger,
    cleanUpLogger:: IO(),
    level :: LogLevels
}

-- initializing logger state by creating a new timed logger (it
takes care of time stamping automatically) and that writes to
stdout. There is a similar function for file logging.

initLogger :: LogLevels -> IO LoggerState
initLogger lvl = do
    timeCache <- newTimeCache simpleTimeFormat
    (logger, cleanUp) <- newTimedFastLogger timeCache
    (LogStdout defaultBufSize)
    let st = LoggerState logger cleanUp lvl
    return st
...

-- src/StackTypes.hs
...
-- type that will hold the State for our App
```

```haskell
data AppState = AppState {
    httpManager :: Manager,
    currentModelId :: Text,
    currentProvider :: ProviderData,
    loggerState :: LoggerState
}
-- Init.hs
...
-- main initialization function in IO to create initial State
and Settings
initAll :: IO (Settings, AppState)
initAll = do
    settings <- initConfig
    let prov = findProviderByName settings "openai"
    print prov
    manager <- newManager tlsManagerSettings
    lgState <- initLogger DEBUG
    let initSt = AppState manager "gpt-4o" prov lgState
    pure (settings, initSt)
```

As you can see from the above, we added loggerState as a separate field in the total application state, and then we adjust the initAll function to make sure the logger state is initialized as well.

Now we can create the logging actions in the Mid monad, which we will use most often.

EXERCISES

1. Change the initAll function above so that it reads the logging level from the .env file instead of being hardcoded.
2. Create logging actions in the Mid monad.

Creating the logging actions in the Mid monad using the functions we already have in IO is straightforward:

```
-- src/Middleware.hs
...
-- LOGGING --
lgFtl :: [Char] -> Mid ()
lgFtl msg = gets loggerState >>= liftIO . lg_ftl msg
lgErr :: [Char] -> Mid ()
lgErr msg = gets loggerState >>= liftIO . lg_err msg
lgWrn :: [Char] -> Mid ()
lgWrn msg = gets loggerState >>= liftIO . lg_wrn msg
lgInf :: [Char] -> Mid ()
lgInf msg = gets loggerState >>= liftIO . lg_inf msg
lgSuc :: [Char] -> Mid ()
lgSuc msg = gets loggerState >>= liftIO . lg_suc msg
lgDbg :: [Char] -> Mid ()
lgDbg msg = gets loggerState >>= liftIO . lg_dbg msg
lgVrb :: [Char] -> Mid ()
lgVrb msg = gets loggerState >>= liftIO . lg_vrb msg
```

We take the loggerState from the overall state and then feed it via a monadic bind operator to the corresponding IO logging action, which is then passed into liftIO – since we are in the Mid monad. Now we have nice shorthands for logging any log level message inside the Mid monad. Of course, you could also write a generic function that takes a log level as an argument, and then you wouldn't need to define a separate function for each level – do this as an exercise.

Finally, let's make sure we are using the version of the logger that writes to the file – simply change the initLogger function to initLoggerFile in the initAll, and all the logs will be written to the ".jarvis_logs" file. We don't want to contaminate the console as we are interacting with Jarvis ourselves, but reviewing the logs later or by running `tail -f .jarvis_logs` in another terminal can be quite handy!

You may ask, "What about the Writer monad? Isn't it good for logging?!" Well, yes and no. It *might* be good for logging, but fast-logger provides a more efficient way to do that, so we don't need to use the Writer monad machinery for this. However, it can be a good exercise.

> **EXERCISE**
>
> Use our logging methods and come up with a way to run logging via the Writer monad. To do it, you will need to come up with a data type that you will need to make an instance of Monoid and run logging functions inside the <> operator. Can you do it without breaking Semigroup and Monoid laws?

Usage Stats in the Writer Monad

Let us now build the Writer part of our RWS monad for something other than logging. OpenAI optionally returns usage statistics in tokens after each request – and this information is a good candidate to summarize at the end of each session, exactly what Writer functionality is good for. There is the following type in the src/LLM/OpenAI.hs:

```
data Usage = Usage
  { prompt_tokens :: Int
  , completion_tokens :: Int
  , total_tokens :: Int
  } deriving (Show, Generic)
```

CHAPTER 10   TERMINAL AI CHAT AGENT

It is exactly what is being returned from OpenAI as JSON and then converted to the respective Haskell data type. All we need to do to be able to use it "automatically" in the Writer part of our monad is make it an instance of Monoid.

> **EXERCISE**
>
> Make Usage an instance of Semigroup and Monoid (for which Semigroup is a prerequisite) by yourself before reading further.

This is a very straightforward exercise by this point, so I am sure you had no trouble writing:

```
instance Semigroup Usage where
  u1 <> u2 = Usage {
    prompt_tokens = prompt_tokens u1 + prompt_tokens u2,
    completion_tokens = completion_tokens u1 + completion_
    tokens u2,
    total_tokens = total_tokens u1 + total_tokens u2
  }

instance Monoid Usage where
  mempty = Usage 0 0 0
```

Semigroup defines a <> operator that combines two elements of our type with associativity, and Monoid adds an empty element to it. Now we can change the definition of our Mid monad as follows: `type Mid = RWST Settings Usage AppState IO`. Instead of a [String] Monoid, we give Usage as a Monoid to the Writer part of the RWS monad, and now we can use the tell function to update the result automatically.

However, before we are able to do that, we need to create a slightly different version of the function that calls OpenAI API and processes the streaming response. First, we need to make it a part of the App monad –

since we will be outputting stuff to the terminal, which is our top floor in the stack. Second, we need to add some better error handling with logging as well as update the stream processing function so that it can be *folded* instead of *mapped*. To do that properly, we need to add the chat history functionality first.

Chat History

To illustrate why keeping track of chat history is important, you can run the program in its state from the previous section and go through the following line of questioning or something similar:

`λInt.` who is Hemingway?
Ernest Hemingway was an influential American author and journalist, born on July 21, 1899, in Oak Park, Illinois, and died on July 2, 1961, in Ketchum, Idaho. He is known for his terse, economical writing style and for pioneering the "iceberg theory" of writing, where the deeper meaning in a story is hidden below the surface.

Hemingway authored numerous classic works of literature, including "The Sun Also Rises," "A Farewell to Arms," "For Whom the Bell Tolls," and "The Old Man and the Sea." He won the Nobel Prize in Literature in 1954, primarily for "The Old Man and the Sea." His experiences as an ambulance driver in World War I, a journalist during the Spanish Civil War, and an adventurer greatly influenced his writing.

Hemingway's lifestyle was as legendary as his prose; he was known for his love of adventure, his passion for hunting, fishing, and bullfighting, and his larger-than-life persona. Despite his external bravado, Hemingway had a complex inner

CHAPTER 10   TERMINAL AI CHAT AGENT

life marked by bouts of depression, and he ultimately took his
own life in 1961. His legacy lives on, with his works remaining
a significant part of the Western literary canon.

λ**Int.** what is his best book?
It's a bit difficult to determine which specific "his" you're
referring to without more context. The "best book" for any
author can be subjective and depends on personal taste and what
kind of themes or writing style you enjoy. Could you please
provide the name of the author or more context so that I can
provide a more accurate recommendation?
λ**Int.**

As you can see from the above, our chat agent has no concept of the previous conversation and treats each message separately. When we ask "what is his best book," we would expect the AI to deduce from the context that we are talking about Hemingway – but it doesn't. So even though the bot in its current state can be quite useful to answer isolated questions, it won't really keep any conversation with you. Let's remedy that!

How about the Writer monad? I mean, all we need to do is to give our RWST transformer [Message] instead of [String] and we can build up the chat history automatically?!

Well, unfortunately, again, no. This looks very tempting – but the Writer monad gives you access to what is inside the "logging" Monoid only after you run the monadic action via runRWST/evalRWST and similar methods – and we only do that at the top level of our program. Of course, we could treat each interaction with the LLM API as a single Writer action and then get the resulting [Message] array as a result – but it would mean we'll anyway need to keep the intermediate message arrays somewhere in the State, so it gets messy.

Another issue is that the context that an LLM can handle is limited, plus the tokens are not free – so ideally you want to give the LLM as much context as needed to perform its functions well, but not too much so that it does not lose attention and you do not overpay. To handle this, we will include a tunable parameter – number of messages to keep in the history to be sent to the LLM with every user request.

To do this, let's update the AppState data type:

```
-- type that will hold the State for our App
data AppState = AppState {
    httpManager :: Manager,
    currentModelId :: Text,
    currentProvider :: ProviderData,
    loggerState :: LoggerState,
    messageHistory :: [Message],
    historySize :: Int
}
```

We have added the messageHistory and historySize parameters – and do not forget to update the initAll function in Init.hs so that it includes the new defaults: `let initSt = AppState manager "gpt-4o" prov lgState [] 6`.

Now let's get to reworking the function that calls OpenAI API. In order for us to achieve the desired result – namely, to be able to update usage data and message history after calling the OpenAI API – we need to change the main functions that work with it. Please review these changes carefully – if you are able to fully understand this code, you will have no issues with anything in Haskell. By now you should be armed enough to be able to do that:

```
-- src/LLM/OpenAI.hs
...
chatCompletion :: [Message] -> Text -> ProviderData -> Maybe ChatOptions -> LoggerState -> (BS.ByteString -> IO (String, Usage)) -> IO (String, Usage)
chatCompletion messages modelId provider opts lgState func = do
```

CHAPTER 10 TERMINAL AI CHAT AGENT

```
    -- Note: parseRequest  throws an exception on invalid URLs,
       so in a real application, you should handle that.
    let url = chatCompletionURL provider
    let key = providerKey provider
    let fopts = Data.Maybe.fromMaybe (providerDefaultOptions
    provider) opts
    lg_dbg ("Calling chatCompletion with URL: " ++ url ++ " and
    key " ++ show key) lgState
    initialRequest <- parseRequest url
    let obj = Data.Aeson.object [
                "model" .= modelId,
                "messages" .= toJSON messages
            ]
    let request = setRequestBodyJSON (combineObjects obj
    (toJSON fopts))
            $ applyBearerAuth key initialRequest

    let postRequest = request { method = "POST" }

    -- Use withResponse to process the streaming response
    httpSink postRequest $ \response -> do
        if statusCode (responseStatus response) /= 200
            then do
                liftIO $ lg_err (show (responseStatus
                response)) lgState
                liftIO $ lg_err (show (responseHeaders
                response)) lgState
                foldMapMC (\x -> do
                                let y = unpack x
                                putStr y
                                pure (y, mempty))
```

```
      else do
        -- Stream the response body to stdout
        foldMapMC func
```

The changes we made here are subtle, but important. First, we are now passing the LoggerState object explicitly to be able to use our logger. We are still in the IO monad, so there is no state, so we need to pass it explicitly. Second, we changed the `mapM_C` function in our response handler to `foldMapMC`, also from the conduit package (the type signature for chatCompletion is updated accordingly). This is a key change. With mapping, we go over all elements of some collection (a list, a stream, a tree, etc.), perform some operation with it, and optionally return a changed collection. With folding, we do the same, but we collect all results in some sort of accumulator value – in this case, it is (`String, Usage`). The foldMapMC function takes care of accumulation for us automatically, thanks to the fact that both String and Usage are a Monoid, so their pair is a Monoid as well! And as we well know by now, we can "fold" Monoids easily using their built-in <> operator – and that's exactly what foldMapMC does for us under the hood. The String in our accumulator is the final message – which we will put into message history – and Usage is, well, usage, which we will put to our Writer monad.

Now the only thing that's left to make sure we have enough low-level API and can fix the rest of the program is to adjust the processResp function so that it returns the (String, Usage) object:

```
processResp :: BS.ByteString -> IO (String, Usage)
processResp ch = do
  foldM
    ( \acc x -> do
        let jsonObject = decode (BL.pack x) :: Maybe
        OpenAIResponse
        case jsonObject of
          Nothing -> return acc
```

```
              Just resp -> do
                if not (null (choices resp)) then do
                  let chs = delta $ head $ choices resp
                  case chs of
                    Nothing -> return acc
                    Just DeltaMessage {content = cnt} -> do
                      let str = T.unpack cnt
                      putStr str
                      return (fst acc ++ str, snd acc)
                else do
                  let u = usage resp
                  case u of
                    Nothing -> return acc
                    Just us -> return (fst acc, snd acc <> us)
    )
    ("", mempty) (processString $ unpack ch)
```

This is not optimized at all. I encourage you to think and optimize it, especially with all the Maybe checks. A slightly annoying complication for this function is that the chunks that OpenAI returns can contain *several* "data:" chunks, so we are forced to do another fold inside our function that's being folded itself. One other way to optimize it is to use some more general functions from the Foldable typeclass instead of foldM. The changes done here are straightforward – we added the (String, Usage) accumulator that's initialized to ("", mempty), and then when there is meaningful input, we update this accumulator either by growing the message string or by adding the usage information.

Now we are ready to write a much nicer-looking high-level interface function in the Mid monad that will utilize this low-level interface changes:

```
-- src/Middleware.hs
...
chatCompletionMid :: Message -> Mid ()
```

```
chatCompletionMid message = do
    st <- get
    let provider = currentProvider st
    let msgs = messageHistory st
    let messages = msgs ++ [message]
    (asMsg, us) <- lift $ chatCompletion messages
      (currentModelId st) provider Nothing (loggerState st)
    processResp
    let messages' = messages ++ [assistantMessage $ pack asMsg]
    modify' (\s -> s {messageHistory = messages'})
    tell us
    pure ()
```

This is a much easier-to-use middleware function that uses our low-level chatCompletion function under the hood. Here we simply get the status, build a message list, then call the low-level function and update the status with new messages, and sum usages using Writer's `tell` function.

Then, making changes to the main application loop, which is also simplified:

```
-- src/App.hs
...
loop :: App ()
loop = do
  lift $ lgInf "starting up Jarvis"
  minput <- HL.getInputLine  (TC.as [TC.bold] "\nλInt. ")
  case minput of
    Nothing -> return ()
    Just ":quit" -> return ()
    -- Just (':':cmds) -> (processCommand (words cmds)) >> loop
    Just input -> lift (chatCompletionMid $ userMessage $
    T.pack input) >> loop
```

So much nicer. Now, let's also make a small change to show the total usage after the user quits the session:

```
main :: IO ()
main = do
    (sett, initSt) <- initAll
    (_, _, w) <- runRWST (runInputT (defaultSettings
{historyFile = Just ".jarvis_history"}) loop) sett initSt
    putStrLn "Total usage:"
    print w
    putStrLn "Bye!"
```

Now we are reading the results of runRWST, ignoring final state but outputting total usage. Try it! stack build, stack exec jarvis. Also feel free to open another terminal and run tail -f .jarvis_logs there to see the logs as well (Figure 10-4).

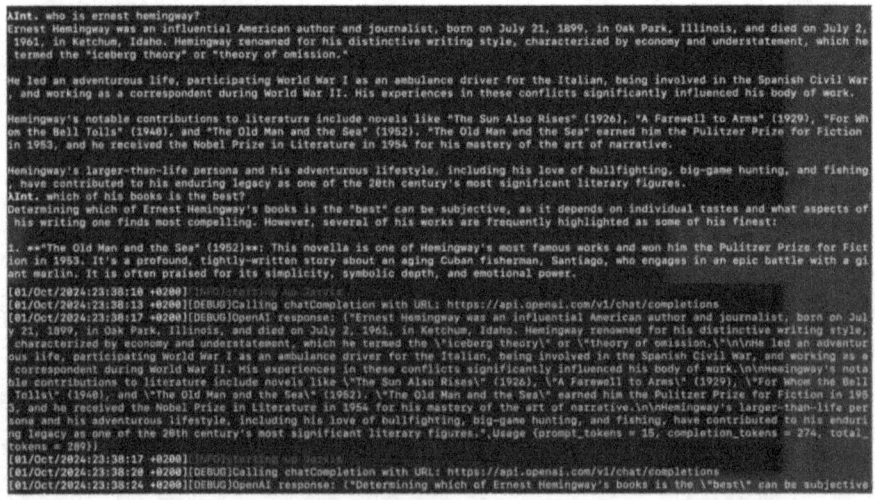

Figure 10-4. *Testing chat history and logging – as you can see, "his books" question is now well understood*

CHAPTER 10   TERMINAL AI CHAT AGENT

> **EXERCISE**
>
> Currently we have not implemented the message history limitation functionality. Do that now – make sure we are only sending no more than N previous messages to the chatCompletion function.

Conclusion

What an amazing job have we done! Constructed a real, "near-production"-quality Haskell program using a monad transformer stack; implemented external API calling and logging; utilized Reader, Writer, and State monads in one go; and again revisited the functional typed design approach. In the next chapters, we will further extend the functionality of our AI framework and will discuss other advanced Haskell concepts, such as mutable variables, vectors, and type families, along the way.

And by the way, you can start using the chatbot we created for help in learning Haskell even now! We will make it even more customizable as a Haskell helper, but even in its current form, you can start asking it Haskell-related questions and it should give you quite good answers. In the next chapter, we will show you some more specific use cases, which can be extremely helpful even for seasoned Haskell programmers – e.g., GHC is known for cryptic error messages, but asking our chatbot to explain them does wonders. Try it yourself or wait until the next chapter, where we will further improve the UI to assist us with such tasks.

CHAPTER 11

Web-Enabled AI Framework and GHC "Guts"

In this chapter, we will further improve "Jarvis" to have better UI and user experience and then will add a basic web access layer on top. This will drive an important discussion on how to use mutable variables in Haskell (it's amazing that we did so much without ever needing them!) as well as a pick under the hood of GHC – for those who are more technically inclined and want to understand how functional machinery works, which may help in optimizing your code for production use.

The code for this chapter is in the "ch11" folder of the book's GitHub repository.

Terminal UI Improvements

Before we dive into the advanced topics listed above, let's review some polishing changes we did to the terminal Jarvis program and discuss a couple of use cases where Jarvis may help you on your Haskell journey specifically. We have added a multiline input mode to the program as

CHAPTER 11   WEB-ENABLED AI FRAMEWORK AND GHC "GUTS"

well as a framework to support whatever commands you may want to put into your UI. You can actually use this as a separate foundation for any terminal-based Haskell program. You can try it by going into the chapter's code folder and doing the usual stack build, stack exec jarvis.

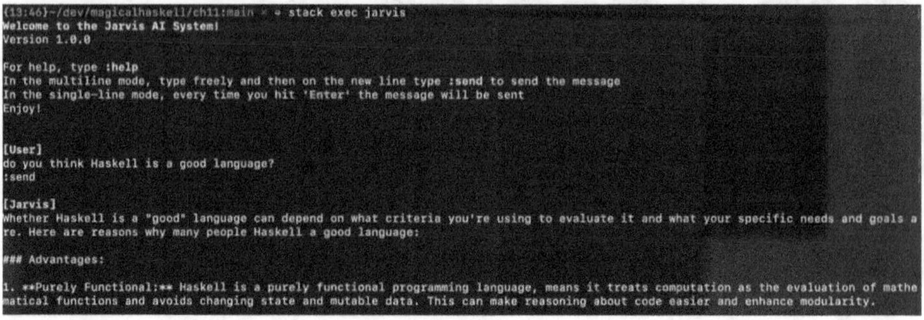

Figure 11-1.   Updated Jarvis terminal UI

To achieve the command support in Jarvis (see Figure 11-1), we updated the AppState a bit with a new object with type UIState:

```
-- src/StackTypes.hs
...
-- various UI settings
data UIState = UIState {
    multilineMode :: Bool, -- multiline entry mode
    currentLineBuffer :: Text -- current buffer if in the
    multiline mode
} deriving (Show)

-- type that will hold the State for our App
data AppState = AppState {
    httpManager :: Manager,
    currentModelId :: Text,
    currentProvider :: ProviderData,
    loggerState :: LoggerState,
```

```
    messageHistory :: [Message],
    historySize :: Int,
    uiState :: UIState
}
```

Currently, we are only tracking if we are in the multiline mode, and then if we are, we are using a buffer for the current text message. I'll show you a couple of cases where we couldn't do anything without the multiline mode – such as asking Jarvis to explain compiler messages or a certain Haskell function. For now, let's finish with reviewing the "command framework":

```
-- src/Commands.hs
import qualified Data.Text as T
import Control.Monad.RWS ( gets, MonadTrans(lift), MonadIO (liftIO), modify' )

import StackTypes
import Middleware (chatCompletionMid, Mid)
import LLM.OpenAI (userMessage)
import System.Console.Haskeline (outputStrLn, InputT)

type App = InputT Mid

data CommandDescription = CommandDescription {
    helpTextShort :: String,
    helpTextLong :: String,
    commandName :: String,
    commandAction :: [String] -> App() -- [String] only includes options
}

allCommands::[CommandDescription]
allCommands = [
```

CHAPTER 11 WEB-ENABLED AI FRAMEWORK AND GHC "GUTS"

```
        CommandDescription {
            helpTextShort = ":send -- send a multiline message
            currently typed",
            helpTextLong = ":send -- send a multiline message
            currently typed",
            commandName = "send",
            commandAction = commandSendText
        },
        CommandDescription {
            helpTextShort = ":multi -- send a multiline message
            currently typed",
            helpTextLong = ":multi -- send a multiline message
            currently typed",
            commandName = "multi",
            commandAction = commandToggleMultiline
        }
    ]
```

processCommand :: [String] -> App()
processCommand ("help":_) = commandHelp
processCommand (cmd:opts) = do
    let c = filter (\x -> commandName x == cmd) allCommands
    if not (null c) then commandAction (head c) opts
    else controlMessage [white, bold] "No such command. Please try :help"
processCommand _ = controlMessage [white, bold] "No such command. Please try :help"

commandHelp :: App()
commandHelp = mapM_ (controlMessage [white,bold] . helpTextShort) allCommands

```haskell
-- toggle multiline mode on or off
commandToggleMultiline :: [String] -> App()
commandToggleMultiline _ = do
    uis <- lift (gets uiState)
    lift $ modify' (\s -> s { uiState = uis {multilineMode =
    not (multilineMode uis)}})
    controlMessage [lgreen] ("[INFO] Multiline mode: " ++ show
    (not (multilineMode uis)))

-- send the text that is currently in the text buffer
commandSendText :: [String] -> App()
commandSendText _ = do
    txt <- currentLineBuffer <$> lift (gets uiState)
    if T.length txt > 0 then do
        (asMsg, _) <- lift (chatCompletionMid $
        userMessage txt)
        uis <- lift (gets uiState)
        lift $ modify' (\s -> s { uiState = uis {
        currentLineBuffer = ""}})
        liftIO $ formatViaLatex (T.pack asMsg)
    else controlMessage [yellow] "[WARNING] Your message is
    empty, please type something first"

-- output ansified text message
controlMessage :: [String] -> String -> App()
controlMessage opts msg = outputStrLn (as opts msg)

-- checking for multiline mode
isMultilineOn :: App Bool
isMultilineOn = multilineMode <$> lift (gets uiState)
```

This module should give you a good idea about how to organize Haskeline-based UI, at least one of the ways. We define a structure (CommandDescription) to describe our commands, each of them is basically an App() action, and then processCommand automatically parses the user's input, and if the line starts with ":", we treat it as a command. It is then forwarded to the correct action to execute. This also allows us to have an "automatic" help function that simply lists all help lines for all functions, and if the user needs more help on a specific function, we can list a "long help line." You get the idea.

Commands `commandToggleMultiline` and `commandSendText` also illustrate how you can manage UI-specific state and its dependencies. In our case we can switch between multi- and single-line modes and depending on the mode either build up the text buffer before sending it to the LLM or send the entered line directly.

We have also added code syntax highlighting and some basic markdown formatting using `cmark` and `skylighting` packages, but we won't go into details on how it is implemented – you have more than enough experience by now to read the code in src/Util/Formatting.hs yourself along with Hackage documentation for these packages to figure it out, and it does not add any structural knowledge.

Now, what can Jarvis be useful for in the context of learning Haskell and programming in general?

Example Jarvis Scenarios

Jarvis can be quite useful to you in various scenarios. First, LLMs are very good in transforming data from one format to another – and this is ubiquitous in programming. Second, the currently used gpt-4o model is trained on Haskell fairly well (even though subjectively Claude 3.5 performs a bit better – but we will discuss how to use other LLMs later in this book) and can give you some directions, even though the code it generates is often with errors, which are not that easy to catch if you are not fluent in Haskell yourself. Having said that, if you give it enough context in terms of the source code you are working with and data types that this source code is using, you can get a pretty good assistant that will help you learn while doing.

CHAPTER 11   WEB-ENABLED AI FRAMEWORK AND GHC "GUTS"

Let's look at some examples. Fire up Jarvis via `stack exec jarvis` and start experimenting!

Format Conversions

As mentioned, this is one of the most common scenarios. For instance, enter Jarvis:

```
[User]
Convert the following JSON format to Haskell datatype:
{
  name: string,
  age: int,
  dob: Date
}
:send
```

The result you get will be similar to Figure 11-2.

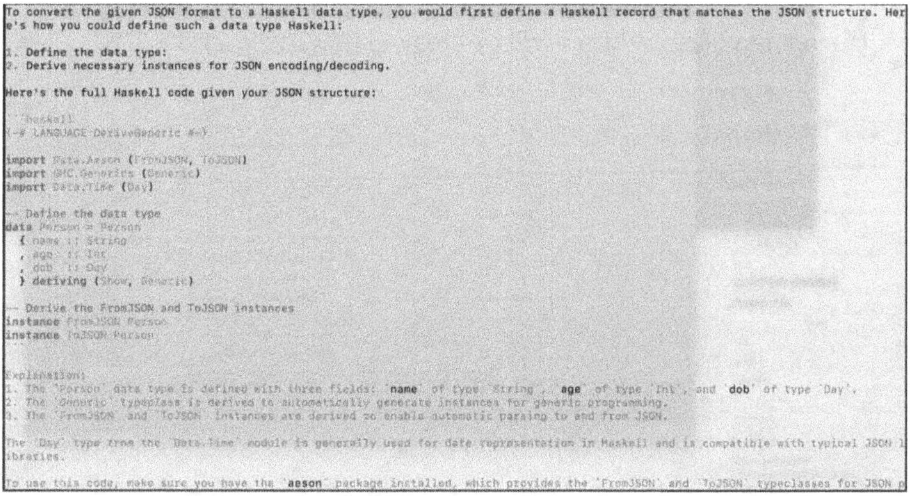

Figure 11-2. JSON-to-Haskell conversion

As you can see, it not only gives you a correct result but provides quite extensive explanations – and you can always ask follow-up questions. There are still quite a lot of mundane "anti-DRY" (do not repeat yourself) tasks when programming, even in haskell, and using AI assistance can make you noticeably faster. And of course this is just the simplest example – but you can also ask to convert unstructured text to JSON or directly into Haskell data types; convert between any formats, especially if you give the AI some examples; etc.

Code Explanation

Let's copy our `loop::App()` code from src/App.hs to Jarvis and ask it to explain this code to us:

```
please explain what the following code does:
<paste code, with imports so that Jarvis knows which libraries we are using>
:send
```

The result you should get will be close to that in Figure 11-3.

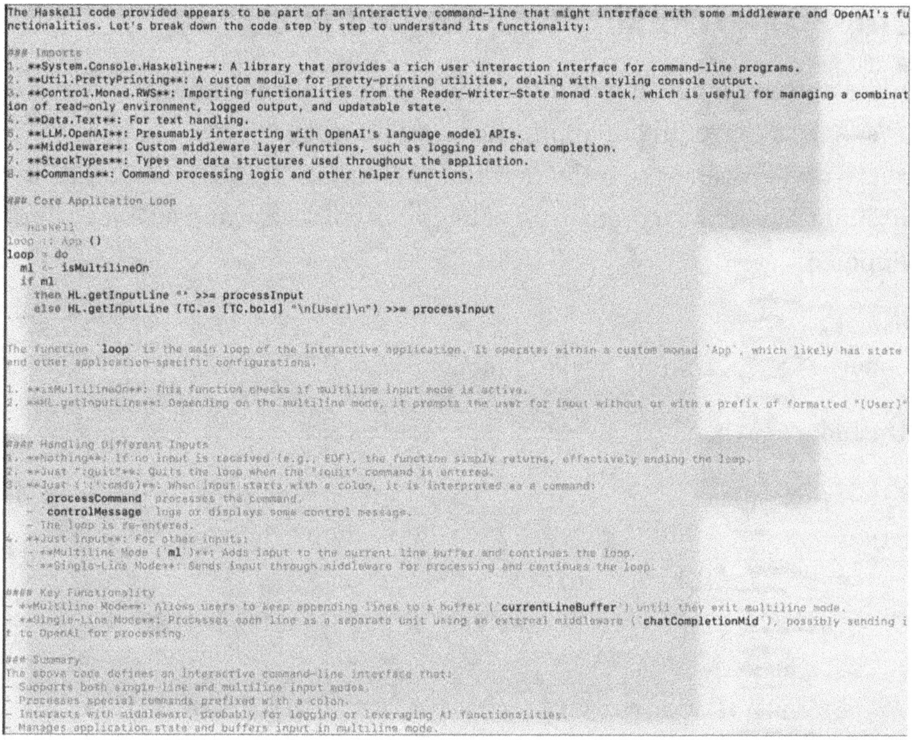

CHAPTER 11 WEB-ENABLED AI FRAMEWORK AND GHC "GUTS"

Figure 11-3. Code explanation

As you can see, the explanations are very detailed, very to the point, and very understandable, as well as the summaries. So, any time you feel lost with some library code, just ask Jarvis to explain it to you, and in most cases this will be quite helpful.

Code generation is a reverse problem, and you can definitely try it as well. Just be mindful that it most likely will be error-prone for now. However, with several iterations working with the compiler, you might be able to get it to work, especially if you make some changes yourself.

CHAPTER 11   WEB-ENABLED AI FRAMEWORK AND GHC "GUTS"

Error Fixing/Working with the Compiler

This is another quick hint that can make you significantly more productive. Let's make an error in our application and then see if Jarvis can help us fix it. We will reverse the order of runInputT and runRWST commands in our main function – a pretty typical beginner mistake – and let's see what happens:

[User]
I have the following haskell code:

<include all headers as well for context!>

```
main :: IO ()
main = do
    (sett, initSt) <- initAll
    evalRWST (lgInf "starting up Jarvis") sett initSt
    showHeader
    putStrLn "" >> putStrLn (as [white,bold] "[User]")
    (_, _, w) <- runInputT (runRWST  (defaultSettings
    {historyFile = Just ".jarvis_history"}) loop) sett initSt
    putStrLn "Total usage:"
    print w
    putStrLn "Bye!"
```

and I get the following errors:

```
/ch11/app/Main.hs:24:29: error:
    • Couldn't match expected type: Settings ((->) StackTypes.
      AppState)
                with actual type: s0 -> m0 (a0, s0, w0)
    • Probable cause: 'runRWST' is applied to too few arguments
      In the first argument of 'runInputT', namely
        '(runRWST
```

```
            (defaultSettings {historyFile = Just ".jarvis_
            history"}) loop)'
    In a stmt of a 'do' block:
      (_, _, w) <- runInputT
                    (runRWST
                        (defaultSettings {historyFile = Just
                        ".jarvis_history"}) loop)
                    sett initSt
    In the expression:
      do (sett, initSt) <- initAll
         evalRWST (lgInf "starting up Jarvis") sett initSt
         showHeader
         putStrLn "" >> putStrLn (as [white, bold] "[User]")
         ....
   |
24 |      (_, _, w) <- runInputT (runRWST  (defaultSettings
{historyFile = Just ".jarvis_history"}) loop) sett initSt
   | ^^^^^^^^^^^^^^^^^^^^^^^^^^^^^^^^^^^^^^^^^^^^^^^^^^^^^^^^
```

<copy all errors, there will be several!>

can you help me fix them?
:send

Here is the result I got, and you should get something similar:

Certainly! Let's go through the issues and correct them step by step.

From the errors, it seems like there's a misunderstanding in the use of the `runRWST` and `runInputT` functions. Let's break it down.

CHAPTER 11 WEB-ENABLED AI FRAMEWORK AND GHC "GUTS"

The `runRWST` function takes a monadic action (of type `RWST r w s m a`), an environment, an initial state and produces a tuple containing the result of the action, the final state, and the writer output.

The `runInputT` function is designed to run a Haskeline input monad, which likely expects a different monadic context.

Given your code, it seems you want to combine an RWST action with Haskeline's `InputT`. However, you need to ensure that `runInputT` wraps around the correct type.

Below is a revised version of your code, correcting the use of `runRWST` and `runInputT`:

<skipping code>

```
-- Use runInputT over an RWST monad action
let settings = defaultSettings { historyFile = Just
".jarvis_history" }
((), _, w) <- runRWST (runInputT settings loop) sett initSt
```

<very nice correction, skipping code..>

Explanation of Changes:
1. `runInputT` now correctly wraps over the `loop` action.
2. `runRWST` wraps around the entire `runInputT settings loop` action, providing the required settings and initial state.
3. Return values from `runRWST` are properly unpacked to handle final writer output.

This should align the types correctly, ensuring your code compiles and runs as expected.

As you can see, Jarvis not only corrected the code but explained very well what the problem was! Frankly speaking, if I had such an assistant when I was learning Haskell, I would have mastered it *much* faster. Use this opportunity!

From here, if your fantasy works similar to mine, I am sure you are already designing an automatic Haskell code writer – so why not make it an advanced exercise.

> **ADVANCED EXERCISE**
>
> Write an AI agent based on Jarvis that generates Haskell code at your request, compiles it using stack, analyzes any error messages, then regenerates code, etc. – until it compiles.

There are lots of other improvements we could put into terminal-based Jarvis to make it even more useful and convenient, but I encourage you to think about those yourself and customize him any way you like. Next, we will expand our Agentic AI framework to support not just terminal but also web access. To do that, we will need to address mutable variables and lightly touch upon multi-
threading as well – and it will be quite different from the beautiful and predictable mathematical world we have been building so far. In order to understand all of these concepts better, here's another "M" subchapter, where we will open the hood of GHC, our beloved Haskell compiler, and look at how it does its job. Feel free to skip this section altogether if you do not want to dig too deep into technicalities, but if you do and if you want to optimize your Haskell programs to the full extent, this might be quite useful for you.

CHAPTER 11 WEB-ENABLED AI FRAMEWORK AND GHC "GUTS"

"M" Subchapter: Under the Hood of GHC

GHC is based on so-called "Spineless Tagless G-machine" (STG) architecture, and the paper by Simon L. Peyton Jones,[1] despite being written in 1992, still reads very well today. The concepts presented there were the foundation of GHC, or Glasgow Haskell Compiler, and despite lots of additions and extensions, the original STG-based architecture still stands.

There are a lot of interesting details in the paper, and I encourage you to read it if you are into low-level details. Here we will discuss only some highlights – as with most of the sections of this book. Haskell is such a vast language that one can write books on each of the concepts we are covering here, so the primary goal here is to give you a strong, practical foundation of how to read the types and design programs in the typed functional way and encourage you to dig deeper yourself. Along with the research papers, of which there are a lot and they are often referenced in the library documentation on Hackage, there is a nice overview of GHC from the *The Architecture of Open Source Applications*[2] book.

So let's look at some important concepts. GHC Haskell is a strongly typed language that uses a so-called "call by need" lazy evaluation approach. GHC has been optimized for performance, and indeed, in many cases related to data processing, it outperforms C/C++ programs, and if you compare with virtual machine–based languages such as Java or C#, this performance gain is even larger. Also, GHC Haskell deals extremely well with concurrency and scales arguably better than any other language with multiple cores. However, to get the most out of these possibilities, it helps to understand the internal machinery.

---

[1] "Implementing lazy functional languages on stock hardware the Spineless Tagless G-machine", Simon L. Peyton Jones
[2] https://aosabook.org/en/v2/ghc.html

First, an important point that a lot of beginners miss. **There is no type information at runtime**. It is completely erased. Types are only used during the compilation stage for typechecking, which guarantees mathematically that our program will run (since all functions compose properly), and then at runtime there are no cycles wasted on typechecking – it is all extremely fast graph reduction.

"G" in STG-machine stands for "graph reduction" – and that is exactly what happens when evaluating the Haskell program. It is being compiled into an intermediary language called "Core," which is a simply typed (conforming to the System FC) lambda calculus. Then a number of optimizations are being run with Core. Then it is being further compiled to STG language – which is itself a very simple untyped lambda calculus and which is described in the STG paper. Then STG is being compiled to so-called "C--" language and then to various backends – machine codes, JavaScript, etc. But the STG structure is kept pretty much directly in these last transformations, so the Haskell program, even compiled, is in fact a lambda calculus in machine codes – very different from an imperative program.

The key issue to understand here is **how data is being represented**. Let's say you have the following ADT sum type and a function to calculate the area:

```
data Shape = Rectangle {
      x :: Float,
      y :: Float,
      name :: String
}
  | Circle {
      r :: Float,
      name :: String
}
```

```
area :: Shape -> Float
area (Rectangle x y _) = x * y
area (Circle r _) = 3.1415926 * r * r
```

At runtime, all data are represented as tuples, which are captured quite well by the pattern match language above. At runtime, the area would be converted to pseudo-Core something like

```
area shape = case (Constructor_of shape) of
                Rectangle -> shape(0) * shape(1)
                Circle -> shape(0) * shape(0) * 3.1415926
```

Constructor tags are still there, even though the tuples are all uniform and look something like 0 (pointer_to_x, pointer_to_y, pointer_to_name), where 0 is the number of the constructor inside our type. So during compilation all type information is erased, apart from constructor tags to distinguish between different product types inside a sum type, and the data is uniformly represented as tuples/thunks together with a code pointer that needs to handle this data.

This gives us the following optimization that is used quite often – the newtype keyword. In Haskell, the newtype keyword is used to define a type that is distinct from an existing type but has the same underlying representation at runtime. It is primarily used for type safety and clarity of code. Types using newtype declarations must have exactly one constructor with exactly one field, and it introduces a new type that is distinct from the type it wraps. This means you must explicitly convert between the new type and the underlying type, which can help prevent certain classes of bugs. Since newtype has only one constructor and one field, it has no runtime overhead compared with the underlying type. The Haskell compiler can optimize away the constructor, so the new type is essentially a zero-cost abstraction.

Suppose you have a type Int that you want to use to represent age. You could define

```haskell
newtype Age = Age Int

-- A function that requires an Age
isValidAge :: Age -> Bool
isValidAge (Age age) = age >= 0 && age <= 120

-- Usage
let age = Age 30
print (isValidAge age)   -- Output: True
```

In this example

- "Age" is a new type that's distinct from "Int."

- You can't directly use an "Int" where an "Age" is expected; you must use the constructor "Age" to wrap the "Int."

- The underlying representation at runtime is just an "Int," without any additional overhead.

The primary purpose of newtype is to provide type safety and expressiveness in your Haskell programs without incurring runtime performance costs. This allows you to use the type system to enforce correctness and clarity in ways that wouldn't be possible if you just reused existing types.

The next point to be aware of is that **Haskell data types are "boxed."** There are so-called primitive types provided to us by the computer architecture – int, float, double, word, byte, etc. – the ones that can be used directly on the CPU. Haskell builds everything on top of this foundation, which allows to treat laziness, optimizations, and all the other "mathematical" features uniformly. However, what it means in practice is that *every* Haskell data type, including seemingly "primitive" Int, is boxed. This means that the tuple of every data type (as in the paragraph above) contains pointers to the heap where the actual data resides. This adds a level of indirection – which is totally okay with complex data types, such as Text, or product types that you build when designing the problem domain – but if you want to run fast mathematical or very low-level byte processing, you are out of luck. Consider this:

```
data Int = I# prim_int
```

This is very close to what an actual Haskell data type Int is defined. It "boxes" the primitive int, so all of your Ints will be represented at runtime as a tuple (pointer_to_int). If you have a vector of such values and map over them, there is one additional operation at each step to dereference the pointer. Definitely not what we want.

Luckily, in many cases GHC optimizes operations with boxed primitive types to unbox them automatically where possible, but if you want more control, you can use unboxed primitive types directly. We won't go into details, but just know that it's possible. Also, there are very good libraries, such as vector – which we will discuss further on – that already optimize everything as needed under the hood and provide amazing performance out of the box (pun unintended).

The last point for this section is **Haskell's laziness and "call by need."** Haskell does not evaluate any expressions until they are needed – unlike most of other languages, which, for instance, evaluate all arguments to a function before calling a function. Haskell doesn't do that; for instance, if at some point of the program you have a situation like this:

```
x = 2*2
...
y = x + 6
```

Haskell will not calculate y! It will be represented as a so-called thunk similar to y = 2*2 + 6, but the calculation itself will not be performed until the program goes to a point where a value is actually *needed*, for instance, print y. This gives Haskell a lot of cool abilities, such as deal with infinite data types (such as lists) – a good abstraction for streams – and so on. Another interesting optimization here is that once x is calculated to be 4 instead of a 2*2 thunk – because it was needed at some place – then it gets updated in place and all future references to x will not be recalculated, but a value of 4 will be used. This gives significant performance improvements (e.g., imperative languages would need to calculate x *every time*). However, sometimes you want to tell Haskell explicitly that you do not want thunks in your data type and it needs to be calculated every time. You can do it using the "!" symbol:

```
data NoThunk = NoThunk {
       age :: !Int,
       name :: Text
}

person = NoThunk 2*2 "John"
-- in the above code, 2*2 will be evaluated immediately thanks
to the "!" sign in the field definition.
```

There are more very interesting architectural decisions in Haskell, such as very light and fast multi-threading, excellent parallelism, advanced concurrent mutability, etc., which are out of scope of this book – but I am sure as you dive deeper into Haskell, you'll research them on your own.

Adding a Web API

With these additional details in our toolbox, we can bravely continue building our AI framework. Let's assume you have created some more complex agents than a simple chatbot that we have made so far – such as the Haskell code-generating agent we discussed in the exercise above – and you want to give access to it via API on the Web. Thanks to our thought-out design using monad transformer stacks, we can reuse all of the

CHAPTER 11   WEB-ENABLED AI FRAMEWORK AND GHC "GUTS"

business logic code we created in pure types and IO and Mid (RWS) monads and simply add a different UI monad on top – instead of Haskeline's InputT, we will use a nice, lightweight, and fast Scotty web framework.

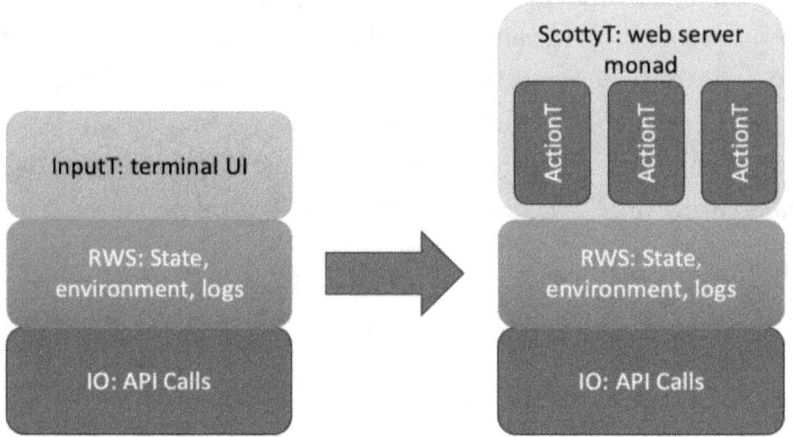

Figure 11-4. *Web API architecture*

As you can see in Figure 11-4, we are keeping our lower floors, as well as pure types and functions, and simply adding another monadic "floor" on top of our stack – and this clean separation is much more difficult to achieve with other languages.

You may also notice that there is not one monad transformer on top now, but two – ScottyT and ActionT. This and some underlying design choices often serve as stumble blocks especially for beginners, so let's look at why we have such a structure in a bit more detail.

A web API is usually structured around different URLs assigned to different actions and different web "verbs": GET, POST, PUT, DELETE. Let us start with just two API calls – generic GET with server status and then POST that allows API users to send a message to our AI agent and receive a response. Later on, as we further enrich our AI framework with vector memory support and the ability to create different agents, this API will become much more useful, but now these two will be enough for the educational purposes.

We will design the following:

- GET request to the /api/v1/status should return a textual status of the server.

- POST request to the /api/v1/chat with a "message" parameter in the request body will call "Jarvis" in the backend and respond with whatever response is given in nicely formatted HTML (including syntax highlighting – since we have it in the terminal, we might as well have it on the Web).

As we have learned in the previous chapters, let's spend some time thinking about the design of our application. So far, we were working with a terminal UI, single-user app, which means it being single-threaded was not an issue. However, web servers are inherently multi-threaded. In fact, warp – the Haskell web server that underlies the Scotty framework – is extremely performant (outperforms nginx on multiple cores by a huge margin[3]) thanks to very lightweight user threads used in the architecture. For all practical purposes, we can think of Scotty/warp creating a new user thread for each new request.

This creates some design problems for us. If we have some settings or state (which we do in our Mid/Reader-Writer-State monad), we can make them available to Scotty actions that will handle certain API requests, but they have no way to update the Mid state – since each action runs in its own thread, how do we pass the pure, updated state from one thread to all the others? It is impossible with what we've learned so far, and the type

---

[3] "The Performance of Open Source Software: Warp," by Kazu Yamamoto, Michael Snoyman, and Andreas Voellmy (https://aosabook.org/en/posa/warp.html)

design of Scotty makes sure to capture this impossibility in a way that is slightly cumbersome and may be confusing for beginners.

Let's look at the main Scotty types and functions – and we will be reviewing the monad transformer versions from the start, since we have already built several monad transformer stacks and it's always better to go from the more general case to much easier bordering-on-trivial "Scotty on top of the IO monad," which you will probably never use anyway.

Scotty provides two monads with corresponding transformers:

```
data ScottyT m a
data ActionT m a
```

The first one is used to define the application as a whole and the second to define specific actions that we want to undertake for each API URL. Then there are functions for each HTTP verb with the same type signature:

```
get  :: MonadUnliftIO m => RoutePattern -> ActionT m () -> ScottyT m ()
post :: MonadUnliftIO m => RoutePattern -> ActionT m () -> ScottyT m ()
-- etc
-- and finally, the usual "run monad" function for Scotty:
scottyT :: (Monad m, MonadIO n)
        => Port
        -> (m W.Response -> IO W.Response) -- ^ Run monad 'm'
           into 'IO', called at each action.
        -> ScottyT m ()
        -> n ()
```

Then creating any web API with the help of Scotty becomes very elegant. We just write something like

```
scottyT port conversion $ do
```

```
get "/" $ text "Hello world in ScottyT"
get "/error" $ do
    status status401
    text "Not authorized!"
post "/api/node/list" $ do
    -- …
    json defaultNodesResponse
-- or in general:
 post url_i action_i
```

Basically, the Scotty web app definition is a list of URLs with corresponding actions to be taken for each depending on the type of the request. Very concise and beautiful, and Scotty does all the inter-monadic conversions under the hood.

The two tricky things that an attentive reader might note are as follows:

- This weird parameter of the scottyT function: (m W.Response -> IO W.Response) -- ^ Run monad 'm' into 'IO', called at each action.

- The type signature of our verb functions that define which action to run for each URL: MonadUnliftIO m => RoutePattern -> ActionT m () -> ScottyT m ()

Let's look at the latter first. It tells us that every "verb"-like function receives a route description and an action to run against it – ActionT m () – and then transforms everything that's happening into "main" Scotty monadic computation – ScottyT m (). So, in our specific case, we need to design an ActionT Mid () function utilizing all the functionality we've developed so far that will take the user's request, go to OpenAI API, and return the nicely formatted response. This means our top application monad will be ScottyT Mid, but it also means that "m" in the type signature above is our monad Mid and it must be a member of a weird new typeclass – `MonadUnliftIO`. As usual, we will search on Hackage and will read the following:

> *While MonadIO allows an IO action to be lifted into another monad, this class captures the opposite concept: allowing you to capture the monadic context. Note that, in order to meet the laws given below, the intuition is that a monad must have no monadic state, but may have monadic context. This essentially limits MonadUnliftIO to ReaderT and IdentityT transformers on top of IO.*

Ouch. Our Mid monad is an RWS, which cannot be made an instance of MonadUnliftIO, since apart from the Reader, it contains Writer and State monads. So what do we do if we need some shared server state between all the actions, some of which may also need to update this same shared state?

We will need mutable variables to keep the global shared state in. Then, our internal actions design can still be nice, mathematical, and monadic – the idea will be that every action that we run in response to get, post, etc. will run in ActionT on top of some Reader monad, which provides the *locally* read-only settings. These settings in turn will have been read from the global mutable variable state but *before* the action starts being executed, so for all practical purposes it will be in the read-only Reader monad while executing. Of course, if an action needs to update the global state, it will also be able to do that, but for us to keep the story as clean and pure as possible, it's best to be done only once at the very end of the action execution. For the schematic overview of this approach, see Figure 11-5.

CHAPTER 11   WEB-ENABLED AI FRAMEWORK AND GHC "GUTS"

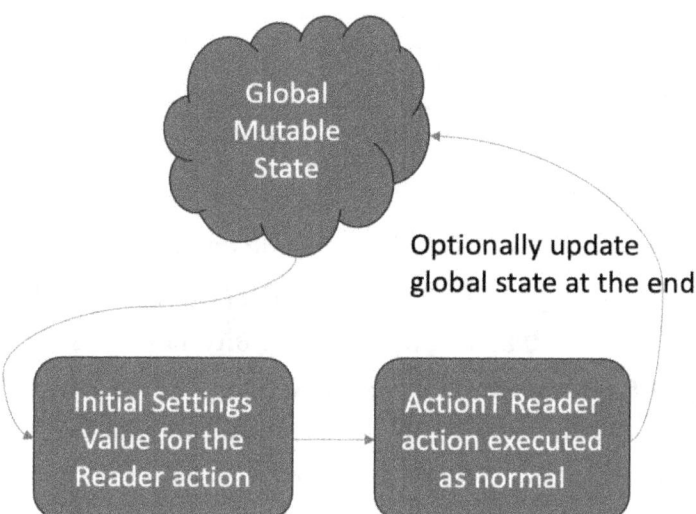

Figure 11-5. Using global state and the Reader monad for Scotty actions

Great, now that we have figured this out, we will use the pattern above to handle state – but we will have to learn about mutable variables first. We will get to it soon enough. For now let's deal with the second weird issue, the "response converter" function that Scotty's runner function expects. It has the following type signature:

```
(m W.Response -> IO W.Response) -- ^ Run monad 'm' into 'IO',
called at each action.
```

This function is run with every request/response and gives us an opportunity to do something with Response in the context of our m monad before converting into IO. We will look at some usage examples down the road.

CHAPTER 11   WEB-ENABLED AI FRAMEWORK AND GHC "GUTS"

> You may think, *Why did we go through the trouble of building an InputT RWST IO stack while designing Jarvis, since now we have to throw away the WS parts of the monad and be stuck with Reader only if we want to do web?*
>
> The answer is that InputT RWST IO and similar stacks are a very good architecture for single-user terminal-based programs and for many other use cases. Web applications are a pretty specific use case that requires *different* design patterns – and this is one of the main goals of this book, to show you versatility of Haskell and teach you to apply different design patterns depending on the task.

Before we start rebuilding our application to properly incorporate it into multi-threaded Scotty design while using global state, let's cheat a little and forcefully make our Mid monad an instance of MonadUnliftIO, so that we can try out our web server in the read-only functionality without any state updates (cheating will be in the fact that we will break monadic laws when making a Mid instance of MonadUnliftIO, which will in essence turn it into a Reader monad – if we make any changes to the State or Writer part of the RWS monad inside ActionT Mid actions, they will be discarded).

Review the code in src/WebAPI/WebApp.hs – and please note that we are cheating as discussed above and will be making significant changes to the Web part as we move along:

```
{-# LANGUAGE OverloadedStrings, DuplicateRecordFields #-}
{-# LANGUAGE TypeSynonymInstances #-}
{-# LANGUAGE FlexibleInstances #-}

module WebAPI.WebApp
where
```

```
import Middleware (Mid, lgDbg)
import Web.Scotty.Trans (ScottyT, scottyT, get, text, status,
ActionT)
import Network.Wai (Response, Request (..))
import StackTypes (AppState, Settings)
import Network.Wai.Handler.Warp (Port)
import Init (initAll)
import Control.Monad.RWS (evalRWST, RWST (..))
import Network.HTTP.Types (status401)
import Data.ByteString.Char8 (unpack)
import Conduit (MonadUnliftIO (withRunInIO))

type WebApp = ScottyT Mid

instance MonadUnliftIO Mid where
    {-# INLINE withRunInIO #-}
    withRunInIO inner = RWST $ \r s ->
        withRunInIO $ \runInIO -> do
            result <- inner $ \action -> do
                (a, s', w) <- runInIO (runRWST action r s)
                return a
            return (result, s, mempty)

tact :: Request -> Mid ()
tact req = do
    let path = rawPathInfo req
    lgDbg $ "Received request on " ++ unpack path ++ ":\n" ++
    show (requestHeaders req)

loggingMiddleware st app req respond = do
    evalRWST (tact req) st
    app req respond
```

CHAPTER 11   WEB-ENABLED AI FRAMEWORK AND GHC "GUTS"

```
mainServer :: Port -> IO ()
mainServer port = do
    (settings, initSt) <- initAll
    scottyT port (convertResponse settings initSt) $ do
            -- middleware (loggingMiddleware st)
            get "/" $ text "Hello world in ScottyT"
            get "/error" $ do
                status status401
                text "Not authorized!"
convertResponse :: Settings -> AppState -> Mid Response -> IO
Response
convertResponse sett st resp = do
    (a, _) <- evalRWST resp sett st
    pure a
```

An instance of MonadUnliftIO code is tricky but not very relevant, so we won't go into it – however, you can decipher if you remember the exercise from one of the early chapters where we made DeckM type instances of Applicative and Monad. Then, the mainServer function defines our Scotty program in exactly the fashion we described above. Inside, we put only two routes to test for status and for the error, and you can check both by running `stack build`, `stack exec web`. This is another neat feature of stack – you can have several executables in the same project, which can be quite handy when you have different interfaces on top of the same framework, just as we are trying to do here. Main for our web program looks as follows:

```
-- src/web/Main.hs
module Main
where

import WebAPI.WebApp

main = mainServer 8080
```

CHAPTER 11   WEB-ENABLED AI FRAMEWORK AND GHC "GUTS"

And we also changed package.yaml to tell stack how to compile it:
executables:

```
jarvis:
  main:                Main.hs
  source-dirs:         app
  ghc-options:
  - -threaded
  - -rtsopts
  - -with-rtsopts=-N
  dependencies:
  - jarvis
web:
  main:                Main.hs
  source-dirs:         web
  ghc-options:
  - -threaded
  - -rtsopts
  - -with-rtsopts=-N
  dependencies:
  - jarvis
```

If you executed stack exec web successfully, you should see "Setting phasers to stun... (port 8080) (ctrl-c to quit)" in the console, and then navigate to localhost:8080 and localhost:8080/error and you should see something similar to Figure 11-6.

CHAPTER 11 WEB-ENABLED AI FRAMEWORK AND GHC "GUTS"

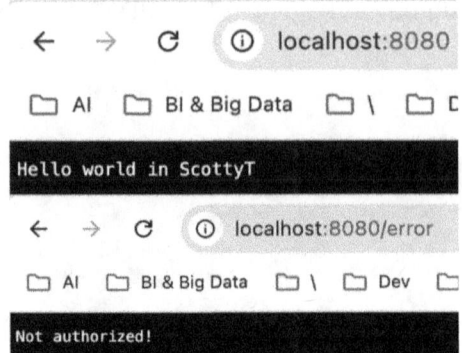

Figure 11-6. *Checking that our basic Scotty server works*

Now we have enough high-level design consideration as well as a very basic practical skeleton of how to run a Scotty web server. The only thing that's left to learn to be able to move from designing single-user terminal applications to multi-threaded web applications with state is how to use mutable variables.

Mutable Variables

We have built some quite advanced programs only using immutable data – and that is always a preferred way to go. Once you introduce mutability, your program becomes prone to various side effects that are difficult to track and debug in the best case, and as soon as you introduce concurrency, there all the usual issues associated with it: race conditions, deadlocks, etc. Even languages such as one of the most terribly designed of all – JavaScript – try to encourage programming using immutable data as much as possible lately. So much more so in Haskell.

However, in some cases it is impossible to do without them, and in some others using mutability in smart ways actually allows creating more performant algorithms while still keeping the guarantees of purity thanks to some advanced typing tricks Haskell provides.

Mutable variables are like the wild beasts of functional programming. In the serene, mathematical world of Haskell, we've been floating alongside pure functions and immutable data structures, but now we're shaking things up a little. Mutable variables are essential when you need to track state between different actions, especially in multi-threaded environments. However, using them is a bit risky, and you should always think twice before reaching for one. The good news is that Haskell provides several powerful abstractions for mutable state, each with specific guarantees and use cases.

We'll explore three of the most commonly used mutable variables in Haskell: IORef, MVar, and TVar. While they appear similar in that they allow mutation, they differ significantly in the level of complexity and control they offer, especially concerning concurrency. Let's dive into each of them, look at their strengths and weaknesses, and figure out when you should reach for each one in your Haskell toolbox.

IORef: Simple Mutable Variables in the IO Monad

An IORef is the most basic form of mutable variable in Haskell. It's a mutable reference that lives in the IO monad, meaning you can read from it or modify it during IO operations. Think of it as a lightweight, thread-unsafe box where you can store (and update) a value, for example:

```
import Data.IORef

-- Basic usage of IORef
main :: IO ()
main = do
    ref <- newIORef 0
    writeIORef ref 42
    val <- readIORef ref
    putStrLn $ "The value in IORef is: " ++ show val
```

In this example, we create an IORef, store the value 42, and then read it back. This is simple enough, but here's the catch: *IORef is not thread-safe.* If multiple threads try to read from and write to the same IORef, you could end up with inconsistent data. This makes IORef ideal for single-threaded applications or for cases where you can guarantee that concurrent access won't happen.

Use Cases for IORef

- Maintaining state in single-threaded applications
- Lightweight counters, accumulators, or caches, where thread safety is not a concern
- Quick and dirty state management in testing or prototyping

When to Avoid IORef

Anytime you expect multiple threads to access the same variable. In such cases, using IORef can lead to race conditions and data corruption.

MVar: Mutable Variables with Concurrency Control

An MVar steps things up by providing more control in concurrent settings. An MVar can be thought of as a box that may either be empty or contain a value. It introduces the concept of synchronization: only one thread can take a value from an MVar at a time. This makes MVar a good candidate when you need locking, signaling, or shared data in a concurrent program:

```
import Control.Concurrent
import Control.Concurrent.MVar
```

```haskell
-- A simple producer-consumer example using MVar
main :: IO ()
main = do
    mvar <- newEmptyMVar
    forkIO $ do
        putMVar mvar "Hello from thread!"  -- Producer
    result <- takeMVar mvar   -- Consumer
    putStrLn result
```

In this example, we create an empty MVar, spawn a new thread that puts a value into the MVar, and then retrieve that value in the main thread. The beauty of MVar is that it ensures safe access between threads. If another thread tries to read from the MVar before the first thread has put a value inside, it will block until the value is ready.

Use Cases for MVar

- Synchronization: Ensuring that only one thread accesses a resource or performs an operation at a time

- Inter-thread communication: Passing values between threads in a safe, concurrent way

- Locking mechanisms: Implementing locks, semaphores, or other synchronization primitives

When to Avoid MVar

- If you have a very complex, high-throughput system where multiple threads need frequent access to shared data, MVar blocking might cause performance bottlenecks.

- If you need compositional concurrency, where multiple transactions or state changes need to be batched atomically.

TVar: The Power of Software Transactional Memory (STM)

Finally, we arrive at TVar, a part of Haskell's STM (Software Transactional Memory) system. TVar takes concurrency control to a whole new level by enabling *composable atomic transactions*. Instead of dealing with locks and blocking like MVar, STM allows you to perform a series of memory operations in a transactional context, ensuring they are all atomic. If there's a conflict (e.g., two transactions modifying the same variable), STM automatically retries the whole transaction:

```
import Control.Concurrent.STM

-- Using TVar in a transactional memory context
main :: IO ()
main = do
    tvar <- atomically $ newTVar 0
    atomically $ modifyTVar' tvar (+10)  -- safely modify inside STM
    val <- atomically $ readTVar tvar
    putStrLn $ "The value in TVar is: " ++ show val
```

In this example, we create a TVar, modify it, and read from it inside atomic blocks. The *atomically* block ensures that all operations within it are either fully completed or fully aborted and retried if there's a conflict.

The STM system shines in complex multi-threaded programs where you need to coordinate changes to multiple TVars. Each `atomically` block guarantees that all changes happen in an isolated, consistent, and durable manner – much like database transactions.

Use Cases for TVar

- Complex concurrent applications, where multiple threads need to read and write shared state.

- Atomic transactions across multiple variables: Imagine a bank account system where you need to atomically transfer money between two accounts.

- Highly composable operations: You can combine multiple transactional functions without worrying about deadlocks or race conditions.

When to Avoid TVar

- If you have very simple concurrency needs (e.g., a single mutable value shared between threads) – in such cases, the overhead of STM may be unnecessary.

- If performance is critical in low-latency systems. STM can be slower than lock-based mechanisms in highly contended environments, although it scales much better than MVar when contention is low.

The Hierarchy Visualized

Table 11-1 summarizes the relationship between these mutable variables.

Table 11-1. Comparison of different types of mutable variables

Feature	IORef	MVar	TVar
Concurrency safe	No	Yes, but blocking	Yes, with transactions; non-blocking
Blocking behavior	None	Yes, can block if empty or full	No blocking; transactions retry automatically
Composability	Minimal	Requires manual composition	Fully composable within STM transactions
Performance	Very fast (single-threaded)	Moderate (depends on blocking)	Context dependent (scales well under contention)
Use cases	Single-threaded state	Shared state, locking, and signaling	Complex concurrent state, atomic transactions

Conclusion

Mutable variables are undeniably powerful but should be used with care in Haskell, as they violate the pristine purity we know and love. That said, IORef, MVar, and TVar each have their own compelling use cases that can make your code more efficient, scalable, and, sometimes, just more *possible*.

Remember that *with great power comes great responsibility*. It's usually a good idea to stick with pure functions and immutable data as long as you can. When you inevitably need to handle concurrency or track mutable state, now you know exactly which tool to reach for, whether it's the straightforward IORef, the synchronization-friendly MVar, or the powerful, composable TVar. Choose wisely, o powerful Haskell wizard!

CHAPTER 11   WEB-ENABLED AI FRAMEWORK AND GHC "GUTS"

EXERCISE

Think about what kind of state you may need in the web API app that serves multiple AI agents to multiple users. Which mutable variable is better suited for the implementation?

In this chapter, we further enhanced the UI of our terminal AI agent and discussed the differences between architecture of the single-threaded terminal-based applications and multi-threaded web applications. We made a quick peek under the hood of GHC to be able to better understand how to optimize its performance. We also finally arrived at the necessity to introduce mutable variables, as handling global state between different threads without them is all but impossible. In the next chapter, we will further design the web API for our AI framework, discuss some enhancements to the agents themselves, and introduce vector-based retrieval augmented generation or "RAG," which significantly improves performance of AI agents, especially if you need them to be able to handle your own, proprietary or personal data. This will drive the discussion of the `vector` library – an extremely efficient and performant library that supports arrays, mutable, as well as immutable. And since it is based on so-called type families – another common obstacle for beginners – we will gently introduce them building up on our typeclass hierarchy.

CHAPTER 12

Down the Rabbit Hole

In this chapter, we will continue building up our AI agents framework using what we've learned and will learn some new advanced concepts along the way. What we want to achieve as an end result now starts to become more specific and real, so let's recap:

- We are creating an Agentic AI framework that allows access via terminal UI and via web.

- We want to share as much code as possible between the two – so this will drive some redesign of what we created so far.

- We need to add some sort of persistence layer – and we will use MongoDB for this purpose. SQL databases are very well covered including in Haskell context, but with the advent of LLMs, unstructured data becomes as important as or even more important than structured. MongoDB is the leader among no-SQL databases, so it makes sense to discuss its usage with Haskell.

- We want to have advanced agent functionality, i.e.:
  - Ability to support multiple agents.
  - Agents should use multiple LLM and other steps, such as integrations with external existing systems.

- We need to be able to have persistent editable "system prompts" for different LLMs, which influence their "tone of voice" and behavior in general.
- We want to support so-called Vector RAG, or retrieval augmented generation.

We shall try to cover and implement as much of the functionality described above in the remaining chapters of this book, and in any case we will at least give specific pointers on how to implement this or that advanced functionality.

In this chapter, we will focus on adding the MongoDB persistence layer, on discussion of vectors and type families in the context of Vector RAG, and on restructuring the architecture of our framework to support both terminal and web API access properly, which will also drive a lot of mutable state–related discussions.

The code for this chapter is in the "ch12-1" folder of the book's GitHub repository for the first part, and then we switch to the "ch12-2" starting with the "Redesigning Architecture: MRWST Monad Transformer" section.

MongoDB as a Persistence Layer

Before we are able to expand our web API functionality with global mutable state, we need to add the persistence layer – since a lot of AI agent improvements depend on it. Please install MongoDB locally using the instructions for your system. On Mac it is easily done with Homebrew and on Linux via package management of your system, and on Windows you'll have to read the documentation on the https://www.mongodb.com/ website.

MongoDB is the leader among so-called "no-SQL" databases. It stores JSON objects more or less directly and as such is very well suited to support web applications, since JSON is ubiquitous in the modern Web. We use it in production at integrail.ai, as well as many other SaaS companies, so learning to use it with Haskell is quite beneficial if you are planning to make production applications as well. The Haskell package that we will use to access Mongo is called mongoDB. As usual, feel free to look it up on Hackage and review the documentation as we go along. The library provides an interface of actions in the IO monad, which makes our lives easier as it gives us flexibility to use those actions from any monad transformer stack we decide to stick on top of the IO.

Before we start building up Mongo functionality, let's make a couple of housekeeping changes that will make our program much more convenient to use: we will add the environment file check during startup sequence to make sure all the needed variables are present, and we will run both the web API and the terminal Jarvis UI in one main file – this will allow us to manage the web server directly from the terminal down the line.

First, don't forget to copy the .env.template file to .env and add the missing variables there. These are the ones that must be filled in at this stage:

```
OPENAI_ORG=
OPENAI_TOKEN=
MAIN_MONGO_URI=
```

•

Make sure your Mongo connection string is in the MAIN_MONGO_URI in the .env file. Then, we extend the src/Init.hs file as follows:

```
-- main initialization function in IO to create initial State
and Settings
initAll :: IO (Settings, AppState)
initAll = do
```

```
    settings <- initConfig
    let prov = findProviderByName settings "openai"
    -- print prov
    manager <- newManager tlsManagerSettings
    lglev <- lookupEnv "LOG_LEVEL"
    let (Just lglev') :: Maybe LogLevels = maybe (Just DEBUG)
    readMaybe lglev
    lgState <- initLoggerFile lglev'
    checkEnvironment lgState
    let initSt = AppState manager "gpt-4o" prov lgState [] 6
    (UIState True "")
    pure (settings, initSt)

checkEnvironmentVar lgs var = do
    e1 <- lookupEnv var
    maybe (lg_err ("Environment variable missing: " ++
    var) lgs)
        (\x -> do lg_suc ("Loaded environment: " ++ var) lgs)
        e1

checkEnvironment lgs = mapM_ (checkEnvironmentVar lgs) [
        "OPENAI_ORG",
        "OPENAI_TOKEN",
        "OPENAI_URL_CHAT",
        "OPENAI_URL_MODELS",
        "OPENAI_URL_EMBEDDINGS",
        "MAIN_MONGO_URI"
    ]
```

We added a simple function that checks for the presence of all the needed variables and logs it, respectively. Eventually, it would be good to add distinction between warnings – for variables that are missing but not critical – and critical errors, for variables that are needed to run properly, and so the program should exit in this case. This is a good exercise for you.

CHAPTER 12 DOWN THE RABBIT HOLE

Finally, let's make some small changes to the main code to run both the web server and the terminal UI in one program:

```
-- app/Main.hs
...
main :: IO ()
main = do
    (sett, initSt) <- initAll
    wid <- forkOS $ mainServer 8080 sett initSt
    evalRWST (lgInf "starting up Jarvis") sett initSt
    showHeader
    putStrLn "" >> putStrLn (as [white,bold] "[User]")
    (_, _, w) <- runRWST (runInputT (defaultSettings
    {historyFile = Just ".jarvis_history"}) loop) sett initSt
...
```

We have added one simple line: `wid <- forkOS $ mainServer 8080 sett initSt` that starts our web server in a separate thread using the Control.Concurrent package. We adjusted the mainServer function slightly to avoid double initialization as well – it's easy to see what it is. There is no need to do any cleanup, as when you exit the main thread (which is the terminal UI in our case), all child threads will be terminated as well. Let's try it: `stack build`, `stack exec jarvis`. Also don't forget to run `tail -f .jarvis_logs` in another terminal window to monitor the logs and navigate to localhost:8080 in the browser to check that both the web server and the terminal UI are working. You should see something similar to Figure 12-1.

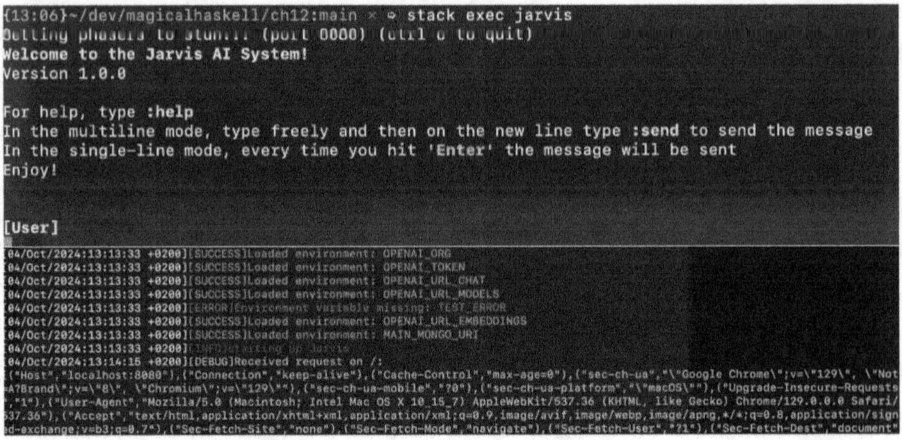

Chapter 12 Down the Rabbit Hole

Figure 12-1. Running Jarvis in both the terminal and web server modes. The top picture is the terminal UI, and the bottom shows logs from both UIs in one place

As you can see, both modes are working fine, and our log file is capturing the logs from both threads as well! Herein lies another problem and an opportunity for an advanced exercise: if both the UI thread and the web app thread will try to write the logs simultaneously, this may lead to an error. Build a proper synchronization mechanism for logging to the file as an exercise.

MongoDB Initialization

Great start! Now let's begin working on the MongoDB functionality. The core Mongo-related types and initialization functions are in the src/Mongo/Core.hs file. Review it at your convenience, it is quite straightforward, and here are some highlights:

```
data MongoConnection = MongoReplica ReplicaSet T.Text
    | MongoPipe Pipe T.Text
...
```

```
initMongo :: LoggerState -> IO MongoState
initMongo lgs = do
    -- we know its there because we checked environment
    (Just muri1) <- lookupEnv "MAIN_MONGO_URI"
    ms1 <- parseMongoStr muri1
    bdb <- connectToMongo ms1 lgs
    return $ MongoState {
        mainConnection = bdb
    }
-- run a mongo action act using our connection conn
withMongoConnection :: MongoConnection -> Action IO a -> IO a
withMongoConnection conn act = do
    pipe <- getPipe conn
    db <- getDatabase conn
    MNG.access pipe master db act
```

The MongoConnection data type stores connections to different MongoDBs, depending on their configuration. initMongo initializes our Mongo-related state, currently consisting of only one connection to the main database, but we will be expanding it along the line. It reads the environment variable for the connection string (note: currently only the "mongodb://..." type of connection strings is supported, not "srv://..."), parses it into a proper connection object, and then actually connects to the database. Of course, there is no connection pooling or functionality for checking for broken connections now – feel free to add it yourself as you are building a production app.

The last function, withMongoConnection, is a convenient wrapper that allows us to execute various MongoDB commands to avoid boilerplate. If you add the initMongo call to the initAll now and if your MongoDB is setup correctly, you should see something like Figure 12-2 in the logs.

CHAPTER 12   DOWN THE RABBIT HOLE

```
[04/Oct/2024:13:59:31 +0200][SUCCESS]Loaded environment: OPENAI_ORG
[04/Oct/2024:13:59:31 +0200][SUCCESS]Loaded environment: OPENAI_TOKEN
[04/Oct/2024:13:59:31 +0200][SUCCESS]Loaded environment: OPENAI_URL_CHAT
[04/Oct/2024:13:59:31 +0200][SUCCESS]Loaded environment: OPENAI_URL_MODELS
[04/Oct/2024:13:59:31 +0200][ERROR]Environment variable missing: TEST_ERROR
[04/Oct/2024:13:59:31 +0200][SUCCESS]Loaded environment: OPENAI_URL_EMBEDDINGS
[04/Oct/2024:13:59:31 +0200][SUCCESS]Loaded environment: MAIN_MONGO_URI
[04/Oct/2024:13:59:31 +0200][INFO]Connecting to localhost
[04/Oct/2024:13:59:31 +0200][SUCCESS]Connected
[04/Oct/2024:13:59:31 +0200][SUCCESS]Mongo Authentication success
[04/Oct/2024:13:59:31 +0200][INFO]starting up Jarvis
```

Figure 12-2. Startup logs with successful MongoDB connection

Adding a System Prompts Collection

To try out our Mongo functionality, let's add one collection that we will be using for sure – the one that will hold our system prompts library, so that we can switch them at will for our AI agents. Create src/Mongo/SystemPrompts.hs as follows:

```
{-# LANGUAGE DeriveGeneric, ScopedTypeVariables,
OverloadedStrings #-}

module Mongo.SystemPrompts
(
    SystemPromptData(..),
    insertSPD,
    findAllSPD
)
where
import Data.Text (Text)
import Data.Time (UTCTime, getCurrentTime)
import GHC.Generics (Generic)
import Data.Aeson (ToJSON (toJSON), FromJSON, Value (Object))
import qualified Util.AesonBson as AB
```

CHAPTER 12 DOWN THE RABBIT HOLE

```haskell
import Database.MongoDB (Collection, insert, Value, find,
Select (select), Cursor)
import Mongo.Core (withMongoConnection, MongoConnection)

data SystemPromptData = SystemPromptData {
    title::Text,
    sysPrompt :: Text,
    createdAt :: UTCTime
} deriving (Show, Generic)

instance ToJSON SystemPromptData
instance FromJSON SystemPromptData

spdColName :: Collection
spdColName = "systemPrompts"

makeSPD :: Text -> Text -> IO SystemPromptData
makeSPD t sp = getCurrentTime >>= \x -> pure
(SystemPromptData t sp x)

insertSPD :: MongoConnection -> Text -> Text -> IO Database.MongoDB.Value
insertSPD conn t sp = do
    spd <- makeSPD t sp
    let (Object jsn) = toJSON spd
    let bsn = AB.bsonifyBound jsn
    withMongoConnection conn $ insert spdColName bsn

-- findAllSPD :: MongoConnection ->
findAllSPD :: MongoConnection -> IO Cursor
findAllSPD conn = withMongoConnection conn $ find (select []
spdColName)
```

CHAPTER 12   DOWN THE RABBIT HOLE

This is a very basic skeleton we'll try to use for every collection in the database. First, we define our "schema" data type – SystemPromptData, very basic, just the title and a system prompt itself. createdAt is optional since Mongo stores the creation time in the object's ID, but there is an option to use Text for _ids so we keep it there just in case. We of course make it an instance of ToJSON/FromJSON, as we will have to convert our objects to JSON. Then there are a couple of helper functions and two interface functions: insertSPD and findAllSPD. As you can see, we are forced to explicitly pass the connection object to each interface function, and we would also want to pass a loggerState to be able to log stuff – so it just screams, "Wrap me in the Reader monad!"

Before we do, which will also lead us to redesign the overall stack and architecture in general a bit – we will need to separate Reader, Writer, and global mutable State (so no more State monad) properly to support both the terminal and the web functionality. Let's give it a quick try and make sure that at least the insertion function works. Put the following lines to the main function:

```
res <- insertSPD (mainConnection $ mongoSettings sett) "Jarvis"
"You are a the best in the world Haskell developer. You enjoy
sharing your knowledge."
print res
```

Build and run Jarvis once. Then don't forget to delete or comment these lines out; otherwise, you will keep inserting this prompt at every start! To check that the document was inserted, log into Mongo shell with your development database and run

```
AtlasLocalDev rs-localdev [direct: primary] studio>
db.systemPrompts.find()
[
  {
    _id: ObjectId('66ffecb4a7f094b789000000'),
```

```
    createdAt: '2024-10-04T13:25:08.308062Z',
    sysPrompt: 'You are a the best in the world Haskell
    developer. You enjoy sharing your knowledge.',
    title: 'Jarvis'
  }
]
```

Super, everything works, but we don't want to expand the MongoDB interface before we update our monad stack properly – this way, our code will be much more concise and elegant. Let's get to it!

Redesigning Architecture: MRWST Monad Transformer

If you recall from the previous chapter, our current architecture looks like Figure 12-3.

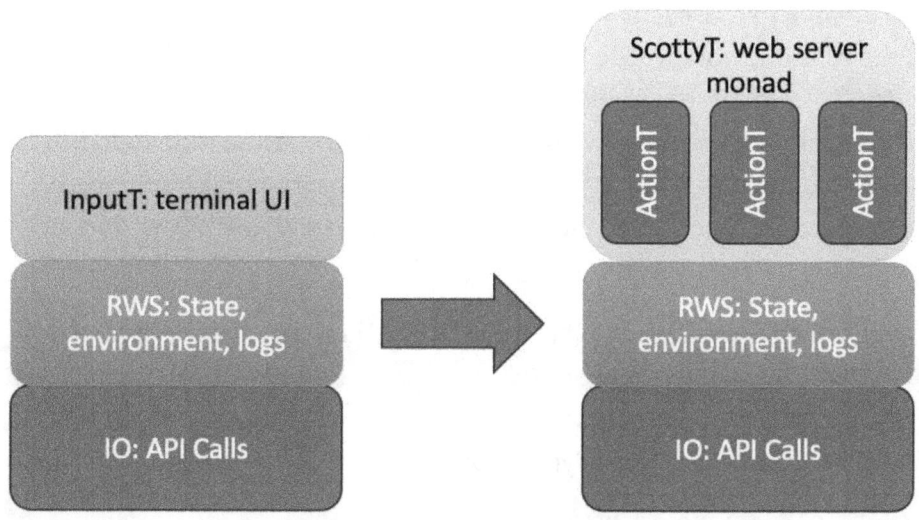

Figure 12-3. Current AI framework architecture

CHAPTER 12   DOWN THE RABBIT HOLE

We need to redesign it to accommodate the persistence layer based on MongoDB, where the Reader monad functionality looks perfect for the task, split our old RWS monad, recombine the new Reader and Writer parts of RWS, and then design global mutable state for multi-threaded access. But maybe we will be able to make it similar to the RWS monad even to avoid inventing new interface functions? If this list seems intimidating, it's only a feeling – all of it is much easier to do than it might sound.

Switch to the "ch12-2" folder of the book's GitHub repository at this point – there are quite a bit of changes to the code structure going forward.

We will take a hint from the Scotty package examples for global state and create a "fake" RWST monad transformer that will in fact be a ReaderT with global mutable state as Reader's settings variable. Note that we will make it a generic monad transformer similar in interface to the RWST – so you are free to reuse it in Scotty-based projects that are fully independent of the AI framework we are building here. Let's call it "mutable RWS," so MRWS(T). Doing so will allow us to make minimal changes to the code we have already created – we will simply change our Mid monad to use MRWST under the hood instead of RWST, and everything should work automagically! The source file is src/Control/Monad/MRWS.hs, and let's go through it step by step:

```
{-# LANGUAGE OverloadedStrings, GeneralizedNewtypeDeriving #-}
{-# LANGUAGE KindSignatures #-}
{-# LANGUAGE FlexibleInstances #-}
{-# LANGUAGE MultiParamTypeClasses #-}

module Control.Monad.MRWS
(MRWST,
runMRWST,
```

```
    evalMRWST,
    execMRWST,
    ask,
    asks,
    tell,
    get,
    gets,
    modify,
    R.lift,
    R.liftIO)
where

import Control.Concurrent.STM
import Control.Monad.IO.Unlift (MonadUnliftIO(..))
import qualified Control.Monad.Reader as R
import Data.Kind (Type)
import Data.Functor ((<&>))
import Data.Bifunctor (Bifunctor(first))
import Conduit (MonadTrans)
import Control.Monad.Catch (MonadMask, MonadCatch, MonadThrow)

-- definition of the MRWST monad
newtype MRWST r w s (m :: Type -> Type) a =
    MRWST { fakeRunMRWST :: R.ReaderT (r, TVar (s,w)) m a }
    deriving (Applicative,
    Functor,
    Monad,
    MonadThrow,
    MonadTrans,
    MonadCatch,
    MonadMask,
    R.MonadIO,
    R.MonadReader (r, TVar (s,w)),
    MonadUnliftIO)
```

CHAPTER 12   DOWN THE RABBIT HOLE

Skipping the import/export parts, we get to our monad transformer definition. Thankfully, we do not have to build it from scratch, but rather we are taking a shortcut and using ReaderT under the hood. The `newtype` definition we discussed in the last chapter serves two purposes here: (1) hides the internal implementation of our monad and (2) makes sure GHC optimizes the runtime code significantly. We are deriving a number of typeclasses that our monad needs to function properly with Scotty and Haskeline monads. Forgetting all the clutter, here is the main part: `MRWST r w s m a = ReaderT (r, TVar (s,w)) m a`. What is happening here?

We are defining a "mutable RWS transformer" with type for the Reader settings `r`, Writer Monoid `w`, and State `s` that runs on top of the monad `m` and returns a value of type `a`. We are doing that by giving an old-and-tried ReaderT a special type `(r, TVar (s,w))`, which defines a pair of read-only `r` and a pair of (`state, writer`), which is in turn put in the mutable TVar box discussed in the last chapter. ReaderT gives us an ability to access the read-only `(r, TVar (s,w))` at any point inside our monadic actions, but since TVar contents can be mutated, we also get the ability to update State `s` and Writer Monoid `w` – and this allows us to build a virtually identical interface to the "regular" RWST monad! Here is the run<monad> function, analogous to runRWST that we used so far in our code:

```
runMRWST :: (R.MonadIO m, Monoid w) =>
    MRWST r w s m a -> r -> s -> m (a, s, w)
runMRWST act settings state = do
    sw <- R.liftIO $ newTVarIO (state, mempty)
    fin <- R.runReaderT (fakeRunMRWST act) (settings, sw)
    sw' <- R.liftIO $ readTVarIO sw
    pure (fin, fst sw', snd sw')
```

The complimentary evalMRWST and execMRWST that return the value and final Monoid (a, w) and state and final Monoid (s, w), respectively, are defined analogously. The runMRWST function gets an MRWST monadic action `act`, read-only Reader `settings`, and initial `state`, and then we do the following steps:

- Initialize a new TVar with an initial state and empty element of our Monoid.

- Unpack our monadic action with fakeRunMRWST to get to the ReaderT contents, and then run the ReaderT monadic action as usual, by using runReaderT act' (settings, sw), where (settings, sw) is exactly the type of (r, TVar (s, w))! Then we bind the resulting value to the fin variable.

- Finally, we read the final values of our TVar "box" – which will contain our final State and final Writer Monoid.

- In the end, we return the (a, s, w) triple in the underlying m monad – just like the "normal" runRWST does.

Think about it and let it sink in. We are completely hiding the mutable machinery inside our monad and making the interface correspond 1:1 with pure, immutable RWST! So, for the end user of our monad, there will be no difference whether they use RWST or MRWST, but the latter will work in multi-threaded environments like a charm automatically.

Now that we have the "run to the underlying monad" functions, we can define (at least parts of) the usual Reader–Writer–State interfaces, with the help of an internal state-modifying function:

```
_modify :: R.MonadIO m => ((s,w) -> (s,w)) -> MRWST r w s m ()
_modify f = R.ask >>= R.liftIO . atomically . flip modifyTVar' f . snd
```

CHAPTER 12   DOWN THE RABBIT HOLE

This function gets a function that modifies our *mutable* state – so only the pair of State and Writer Monoid – and modifies it by the following steps:

- R.ask is a "free" Reader interface ask function that we get automatically for our monad. We use it to get access to the (r, TVar(s, w)) pair.
- The result is being fed to snd (you have to read the functions using the point-free . operator right to left) to get to just TVar (s,w).
- flip modifyTVar' f modifies our TVar using the supplied pure function; flip is needed to make the order of arguments right.
- atomically was discussed in the last chapter; it is needed to handle concurrency issues in the STM monad that gives us TVar.
- liftIO is our usual lifting of the IO action (and all operations with TVar are in IO now) into our MRWST monad.

Beautiful, isn't it? Now we can define the interface functions:

```
-- READER INTERFACE is semi-AUTOMATIC thanks to deriving MonadReader --
-- But we need to extract and manipulate with data a bit to make it
-- "really" like Reader:
ask :: Monad m => MRWST r w s m r
ask = R.asks fst

asks :: Monad m => (r -> a) -> MRWST r w s m a
asks sel = ask <&> sel
```

```
-- WRITER INTERFACE --
tell :: (R.MonadIO m, Semigroup w) => w -> MRWST r w s m ()
tell wrt = _modify (\(s,w) -> (s, w <> wrt))

-- STATE INTERFACE --
get :: R.MonadIO m => MRWST r w s m s
get = (R.ask >>= R.liftIO . readTVarIO . snd) <&> fst

gets :: R.MonadIO m => (s -> a) -> MRWST r w s m a
gets sel = sel <$> get

modify :: R.MonadIO m => (s -> s) -> MRWST r w s m ()
modify func = _modify (first func)
```

The above should be easy to follow for you by now. I encourage you to go through these functions line by line and look up functions and operators that are new for you – if you carefully follow the types, as is our usual way in Haskell, you will understand everything and will start getting an ability to write very concise Haskell code under your belt. For instance, `<&>` and `first` may be of note (don't confuse with `fst`). Here we have only defined parts of the RWST interface that we've used so far in our program, but now the real magic happens. All we need to do is redefine our Mid monad to use MRWST instead of RWST:

```
type Mid = MRWST Settings Usage AppState IO
```

Of course, you also need to update the imports – instead of RWS, import Control.Monad.MRWS that we have just defined above – and then update the runRWST functions to runMRWST and same for eval/exec.

That's it. It takes literally three to four changed lines, and all the work we have done in the previous chapters is not wasted, but rather upgraded to use our new cool mutable RWST monad. Figure 12-4 is an updated architecture schematic showing our new design.

CHAPTER 12   DOWN THE RABBIT HOLE

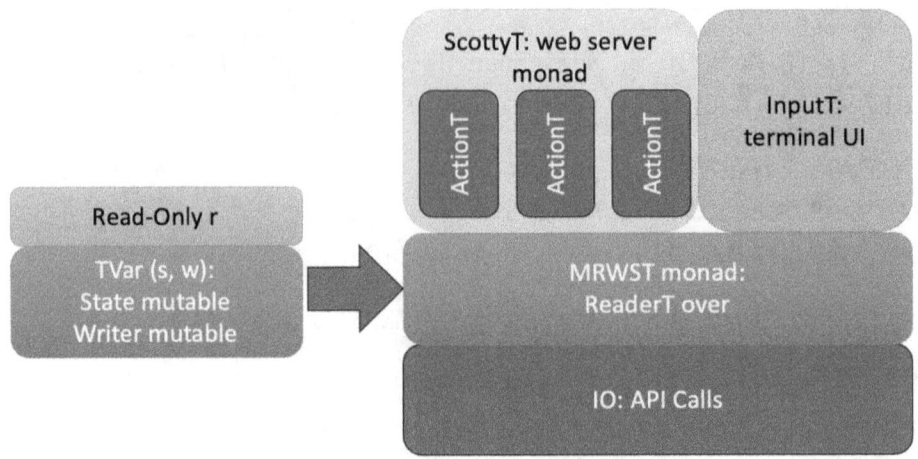

Figure 12-4. *Our final architecture*

By substituting the "old" RWST monad for our new MRWST, we get an excellent benefit of reusing all the code we have created so far and the ability to use both Scotty and Haskeline in one program, running in different threads. MRWST in turn uses a simple ReaderT under the hood, but this ReaderT gives us access to a complex (r, TVar (s,w)) data type shown on the left of the picture. Here, r stays pure and read-only by design, while we get an opportunity to update State and the Writer Monoid thanks to them being put inside a TVar "box." At the same time, all the interface for users of the MRWST monad remains "pure"; there is no mention of mutability, threads, or TVar boxes!

If this doesn't give you goosebumps, I don't know what will.

EXERCISE

Define the remaining standard interface functions from Reader, Writer, and State for our MRWST monad.

Now that we have updated our architectural skeleton, we can continue building up muscles.

Vectors and Type Families

Type families may seem confusing for Haskell beginners at first, but they further extend the power of typeclasses to give us unparalleled flexibility in designing data types and interfaces. Since we are going to need vectors from an amazing vector library next, and they are built both using type families and advanced mutability concepts under the hood, let us try to understand type families just as gently as we did with typeclasses in the very early chapters of the book.

Recall Chapter 3, where we discussed four kinds of functions possible in Type Theory:

- "Normal functions": a -> b in Haskell. Take a value of type a as an argument and return a value of type b as a result.

- "Type functions": Type -> Type in Haskell. Take some concrete type as an argument and return another concrete type as a result. The main mechanism to construct new algebraic data types (ADTs) in Haskell.

- Functions from types to values: Type -> a in Haskell syntax. Take some concrete type as an argument and return a value of some type a as a result.

- Functions from values to types: a -> Type in Haskell syntax. This is what people normally understand by "dependent types" and what is again strictly speaking not *fully* available in Haskell, but there is very active research and some GHC extensions allow some types (such as String or Nat) to be used in limited dependent functions.

CHAPTER 12   DOWN THE RABBIT HOLE

Type Synonym Families

Let's consider the "type functions" first. We have said that our regular data types defined similar to data Maybe a = Just a | Nothing are in fact type functions, and we can rewrite this definition as Maybe (a : Type) = Just x : a + Nothing, which follows closely Haskell's own GADT notation:

```
data Maybe a where
    Nothing :: Maybe a
    Just    :: a -> Maybe a
```

This tells us that Maybe is a function that takes any type "a" as a parameter and creates a new type Maybe a out of it with the help of two constructor functions – Nothing and Just. This is a primary way to create new data types in Haskell. Maybe here has a type Type -> Type – it takes a type and creates a new type.

However, what if we want to not create new types, but have also a kind of "type functions," which return *existing* types? For example, what if I want to have a function like this:

```
AddTypes :: Type -> Type -> Type
AddTypes Int Double = Double
AddTypes Int Float = Float
AddTypes Int String = Bool
```

This is a contrived example obviously, but the point is what if I want to do some type-level manipulations with types to calculate some type value dependent on other types? Well, turns out we can with **type synonym families**:

```
-- src/Misc/TypeFam.hs
{-# LANGUAGE TypeFamilies, MultiParamTypeClasses #-}

module Misc.TypeFam
where
import Data.Kind (Type)
```

```
type family AddTypes a b :: Type
type instance AddTypes Int Double = Double
type instance AddTypes Int String = Bool

-- then we can write:
func :: String -> AddTypes Int String
func "hello" = True -- OK, it is Bool
```

As you can see, `type instance` basically defines the type-level function that we want in the example above. Once the instances are defined, we can use them in function signatures. This *might* be useful when you want to have different types of some data structure depending on which types are used to create it or something, but frankly for practical purposes, I recommend to always start with associated type and data families, which we will discuss further on.

Data Families

Okay, so type synonym families enable us to perform some basic type-level calculations and manipulations. Let us now consider a poster "type function" from the first part of the book, List: `data List a = Cons a (List a) | Nil`. This is a "type function" similar to Maybe above – it takes a type and creates another type from it, but not just that, it tells us *how to construct values of this type* through the constructor functions. However, this is a somewhat limited approach since we define *the same* internal structure for *all types* in the universe.

What if we want to have different internal representations of the List data structure depending on what we are storing in this list? For example, I can represent the list of bits as a structure above, but it will be extremely inefficient – I will be much better off representing a list of bits as a sequence of unboxed (see the previous chapter, section "'M' Subchapter:

CHAPTER 12   DOWN THE RABBIT HOLE

Under the Hood of GHC") bytes, but I still want to treat it as a "normal" list, i.e., to be able to apply all my versatile List functions. Turns out, with **data families** it is possible:

```
data family SuperList a :: Type
data instance SuperList Char = SCons !Char !(SuperList Char) | SNil
newtype instance SuperList () = SListUnit Int
type Bit = Bool
data instance SuperList Bit = SBitList [Word8]
```

Okay, this is pretty cool. Now what if I want to define some familiar List functions for this family, such as length and map? It is possible, but gets tedious, as we need to define everything via typeclasses:

```
class SLength a where
    slength :: SuperList a -> Int

class SMap a b where
    smap :: (a -> b) -> SuperList a -> SuperList b

instance SMap Char Char where
    smap _ SNil = SNil
    smap f (SCons c cs) = SCons (f c) (smap f cs)

instance SLength Char where
    slength SNil = 0
    slength (SCons _ xs) = 1 + slength xs

-- Instance for SuperList () (unmappable, no meaningful transformation)
instance SMap () () where
    smap _ (SListUnit n) = SListUnit n
```

CHAPTER 12   DOWN THE RABBIT HOLE

```
instance SLength () where
    slength (SListUnit n) = n

instance SMap Char () where
    smap _ lst = SListUnit (slength lst)
```

As you can see, even though length is fairly straightforward, when we start writing map, we are forced to write it for every possible pair of the "a" types our SuperList is defined for or leave some cases undefined – which is not ideal for any good library. So even though data families provide some additional flexibility, there are trade-offs you have to think about.

Type synonym and data families also allow you to do some fairly advanced type manipulations with various *DataKinds* extensions of GHC, but this is a very advanced topic out of scope of this book.

Associated Type and Data Families

Let's revisit Chapter 3 a bit more. There, we have sort of formally introduced so-called sigma types and typeclasses, single- and multiparameter. We noted that single-parameter typeclasses such as Monoid add internal structure to our types, while multiparameter typeclasses add structure to the whole *category* of our program by defining relations between different types (see Figure 12-5 as an illustration).

CHAPTER 12   DOWN THE RABBIT HOLE

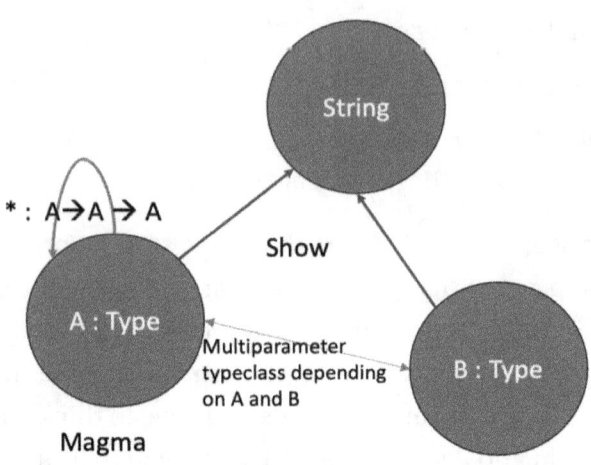

Figure 12-5. Category of a simple program and relations between types defined by typeclasses

We have defined a bunch of typeclasses practically in the first part of the book, so you should feel very comfortable with this concept. Now let's forget Haskell for a minute and just do some "mathematical" write-ups of what typeclasses are. We won't be using actual "Sigma" signs that we used in Chapter 3 as the notation gets cluttered; instead let's invent our own notation and write a couple of familiar typeclasses in it:

```
Show = Sigma (a : Type) {
      show : a -> String
}
Monoid = Sigma (a: Type) {
      (<>) : a -> a -> a,
      mempty : a,
      some laws which cannot be formally written in haskell
}
Functor = Sigma (f : Type -> Type) {
      fmap : (a -> b) -> f a -> f b,
      some laws...
}
```

CHAPTER 12 DOWN THE RABBIT HOLE

```
Monad = Sigma (m : Type -> Type) {
      (>>=) : (a -> m b) -> m a -> m b,
      pure : a -> m a,
      some laws ...
}
SomeMulticlass = Sigma (a: Type, b : Type) {
      someFunction : a -> b
}
```

You get the idea. Every typeclass has a number of variables it depends on (only type variables in the case of Haskell; they are listed in parentheses after Sigma) and then a list of functions that define some sort of structure for our types or relationships between types in case of multiparameter typeclasses. But even in the very first chapter we saw that there are "normal" functions between values and then there are "type functions" – which create new types from other types, such as List a, Maybe a, etc. Well, why are we limiting ourselves? How about including not just "normal" functions but also "type functions" in the list of functions that define our typeclass? I want to write something like

```
MyClass = Sigma (a : Type) {
      -- we can write elem : a, so why not Type?
      SomeType : Type
      SomeTypeFunction : Type -> Type
      func1 : someType -> a
      func2 : someType -> SomeTypeFunction a
}
```

This way, I am introducing in the Sigma "interface" not just "normal" functions that depend on a : Type, but also new types and type functions. This obviously gives us additional flexibility, as we can define not just an actual interface to work with our a : Type, but also some data types, which will have *different internal representation depending on which a : Type is used*! But isn't it just what we saw in the previous two sections, only put inside a typeclass?

Exactly – this is what's called **associated type and data families** in Haskell. I would argue that if you want to use type and data families at all, you are almost always better off using the associated version. It tends to produce much clearer and more maintainable code, which is very modular, as opposed to the "normal" type synonym and data families we looked at in the previous two sections.

Where can they be useful? Apart from the ubiquitous examples of various collections, how about type-safe database representations?

```
-- src/Misc/TypeFam.hs
{-# LANGUAGE TypeFamilies, MultiParamTypeClasses #-}
{-# LANGUAGE DataKinds #-}
import GHC.TypeLits (Symbol)
...
class Database a where
        type TableName a :: Symbol
        type Record a :: Type

data User = User { userId :: Int, userName :: String }

instance Database User where
    type TableName User = "users"
    type Record User = User
```

The above is an example of an associated type family. You need to enable the DataKinds extension for it to work, and then suddenly as you can see we are able to assign a type synonym to a string – in effect making the typeclass dependent not just on *types*, but on *values*! This is way beyond the standard Haskell and one of the cool GHC extensions that you may want to experiment with as you dig deeper.

For an interesting associated data family example, how about a uniform file system interface regardless of how and where it is located?

```
class FileSystem fs where
    data FileHandle fs
    data MyFilePath fs
    openFile :: MyFilePath fs -> IO (FileHandle fs)
    closeFile :: FileHandle fs -> IO ()

data LocalFS
data CloudFS

instance FileSystem LocalFS where
    data FileHandle LocalFS = LocalFileHandle Int  -- File descriptor
    data MyFilePath LocalFS = LocalPath String
    openFile (LocalPath path) = putStrLn ("Opening local file at " ++ path) >> return (LocalFileHandle 1)
    closeFile (LocalFileHandle fd) = putStrLn ("Closing local file with descriptor " ++ show fd)

instance FileSystem CloudFS where
    data FileHandle CloudFS = CloudFileHandle Int
    data MyFilePath CloudFS = CloudPath String
    openFile (CloudPath path) = putStrLn ("Opening cloud file at " ++ path) >> return (CloudFileHandle 1)
    closeFile (CloudFileHandle fd) = putStrLn ("Closing cloud file with descriptor " ++ show fd)
```

The above code is of course very simplified, but a similar approach can be taken to a lot of real-world cases that require different underlying representations of data while being "the same" concept in general – so that we would want to use a uniform interface to access and manipulate it.

> **EXERCISE**
>
> Come up with at least two vexamples of real-world problems where associated type families and associated data families can be a good and elegant solution.

As by now I hope you can see there is nothing complex in the type and data families in Haskell. It's just that nomenclature and semantics of their implementation are somewhat confusing – from a type-theoretical point of view, standalone type and data families would simply be a type of type-level functions, potentially dependent on values as well, while associated families simply a type of sigma data types, just as typeclasses. But once you experiment with them a bit, they will become just as natural as typeclasses have become by now.

Vectors As Arrays in Haskell

Now we know enough to start using the vector library. As usual, feel free to search for it on Hackage and review the documentation as you are reading the book. Vectors in haskell are an answer to the question "what about arrays." Lists are fine and ubiquitous, but not very performant. In many cases, especially when we work with numerical data, we want to be much faster. Also, we want to have mutable arrays in some cases – e.g., for database-like tasks, where performance is key as well.

All of these issues are addressed by the vector library. In fact, Haskell vectors' performance is sometimes better than that of the low-level C code![1]

---

[1] See, for instance, this link: https://medium.com/superstringtheory/once-more-on-haskell-vs-c-performance-a54498bfa91f

Understanding Vectors: Boxed vs. Unboxed

At the heart of the "vector" library lies the differentiation between *boxed* and *unboxed* vectors. This distinction is crucial for understanding how the vectors achieve their remarkable efficiency and for choosing the right type of vector for your application.

- Boxed vectors: These are akin to Haskell's lists but optimized for performance. Each element in a boxed vector is a reference to a value, allowing for heterogenous data types and lazy evaluation. However, because of the indirection involved, boxed vectors aren't the most efficient way to store primitive data types like numbers.

- Unboxed vectors: In contrast, unboxed vectors store elements contiguously in memory, without any indirection. This makes them both time and space efficient for storing primitive types, as they make better use of CPU cache and reduce garbage collection overhead. The trade-off, however, is that they only support types that have instances of the Unbox typeclass, essentially restricting you to a fixed set of primitive data types.

Internal Implementation and Generic Interface

The vector library's internal implementation is finely tuned to balance flexibility with performance. This is achieved by leveraging Haskell's powerful type system and GHC extensions. At a high level, the library provides a generic interface that abstracts over the boxed/unboxed dichotomy. This means you can write polymorphic functions that work with any kind of vector, leaving the specificity to the type inference engine.

Here's a snippet that showcases the generic interface:

```
import qualified Data.Vector as V
import qualified Data.Vector.Generic as GV
import qualified Data.Vector.Unboxed as UV

squareElements :: (Num a, GV.Vector v a) => v a -> v a
squareElements = GV.map (^2)

boxedVecExample :: V.Vector Int
boxedVecExample = squareElements (V.fromList [1, 2, 3, 4])

unboxedVecExample :: UV.Vector Int
unboxedVecExample = squareElements (UV.fromList [1, 2, 3, 4])
```

In this example, `squareElements` is a function that works with both boxed and unboxed vectors, showcasing the generic interface's power and elegance. This is achieved by using the mechanism of type families we just reviewed – boxed and unboxed vectors are represented internally differently, but there is the same interface available.

Mutable Vectors and the ST Monad: For When Performance Is Key

A standout feature of the vector library is its support for mutable vectors, found in the `Data.Vector.Mutable` and `Data.Vector.Unboxed.Mutable` modules. Mutable vectors provide imperative-style, in-place updates, enabling a significant performance boost in scenarios where repeated transformations or aggregations are needed – think numerical simulations or real-time data processing.

Here's a basic illustration of using mutable vectors:

```
import qualified Data.Vector.Mutable as MV
import Control.Monad.ST (runST)
```

```
incrementVector :: Num a => [a] -> [a]
incrementVector lst = runST $ do
    -- Convert the list to a mutable vector
  mVec <- MV.new (length lst)
  -- Write the input list into the mutable vector
  forM_ (zip [0..] lst) $ \(i, value) -> MV.write mVec i value
  -- Increment each element in the mutable vector
  forM_ [0..(length lst - 1)] $ \i -> do
    val <- MV.read mVec i
    MV.write mVec i (val + 1)
  -- Freeze the mutable vector to make it immutable and then
     convert to a list
  frozenVec <- V.freeze mVec
  return (V.toList frozenVec)
```

The incrementVector function demonstrates a common pattern: create a mutable vector, perform in-place updates, and convert it back to an immutable state. While this imperative approach may initially feel unfamiliar in the Haskell context, it's an invaluable tool in optimizing performance-sensitive applications. What is happening here is an ST monadic action (hence runST, more on it below) where we

- Create a mutable vector of the length of the incoming list.

- Copy the list to the mutable vector.

- Increment the values of the mutable vector in place, so no new memory allocations are happening – a very much "side-effectful" operation.

- Then we "freeze" the vector to make it immutable, convert it to a list, and return as a result.

Of course all of these operations are possible without converting back and forth from lists working just with vectors.

An attentive reader might wonder, how in the world do we do a typical side-effectful "IO" operation – mutating memory – but then escape back into the pure world? There is no "IO" or another monad on top of the resulting list. We get it back nice and absolutely pure! How is that possible?

This is the magic of the ST monad (do not confuse with STM, which is Software Transactional Memory from where we get TVar, or with the State monad, which is a "normal" pure monad that we have used previously a lot). If you look at the Hackage documentation, you will read that shown in Figure 12-6.

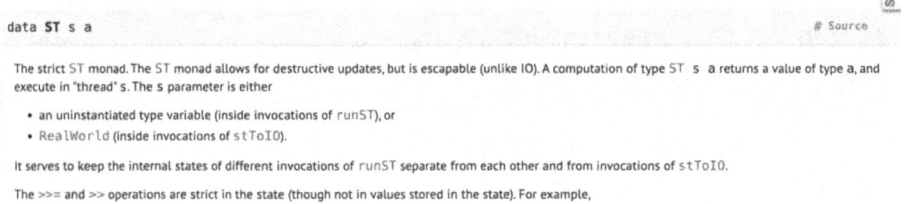

Figure 12-6. *ST monad*

This is a very advanced topic as the ST monad uses so-called rank-2 typing under the hood, but here is a brief explanation. It should be enough for you in practice, and if you want to learn more about higher rank types in Haskell, you can investigate separately.

You do not need to understand the internal workings of the ST monad or what rank-N types are to use them efficiently in practice. The below is for the inquisitive minds among you.

The ST monad in Haskell is a powerful construct that allows you to perform mutable operations safely within a local scope. Let's break down the type signature of runST and how the ST monad works, to understand its unique features and benefits.

The type signature of `runST` is `runST :: (forall s. ST s a) -> a`. Here's how to interpret this:

- ST s a: This represents a computation that performs mutable operations and eventually returns a value of type a. The s parameter is a type variable that is used to ensure the computations in the ST monad remain isolated and safe. It acts as a "phantom type," which prevents you from mixing mutable states across different runST invocations. Basically, it's a typing mechanism that makes sure you never address the same memory space from two different runST calls.

- forall s.: This part is crucial and defines a 2nd rank type (see a simpler example below). It indicates that the computation is polymorphic over the state thread s. This means you can't leak internal mutable state outside the scope of runST, because each call to runST will get its own isolated state.

- a: The result returned by the runST function after executing the ST computation.

Here's a step-by-step explanation of what's happening with `runST` and how the ST monad is used:

- Isolation of mutable state: The ST monad allows you to interact with mutable values safely within a controlled environment. Unlike the IO monad, which allows side effects that can propagate globally, the ST monad confines effects to the local computation within runST and guarantees that side effects do not "leak" into the pure Haskell world.

- Type safety and purity: By leveraging the forall s. quantification, the ST computation ensures that no mutable state can escape its local scope. This is enforced by the Haskell type system; you can't return mutable references or states created inside ST computations because they depend on the universally quantified s.

- Efficiency: Though the operations look pure from the outside, they can be executed efficiently using in-place updates internally, akin to typical imperative programming. This allows for high-performance computations when manipulating complex data structures.

- Transformation to pure values: The computation in ST s a transforms to a pure value a upon completion. The mutable operations are abstracted away, presenting users only with a pure API.

Here's another simple example showcasing runST and the use of mutable references:

```
import Control.Monad.ST
import Data.STRef

sumST :: Num a => [a] -> a
sumST xs = runST $ do
  sumRef <- newSTRef 0    -- Create a new mutable reference
  mapM_ (modifySTRef sumRef . (+)) xs
  readSTRef sumRef    -- Return the pure result

result :: Int
result = sumST [1, 2, 3, 4, 5]   -- usage
```

- **newSTRef** is used to create a new mutable reference inside the ST computation.
- **modifySTRef** modifies the value of an STRef.
- **readSTRef** retrieves the value from a mutable reference.

The above computation sums a list of numbers purely, but achieves internal efficiency with mutable operations by doing it "in place." It's a testament to Haskell's ability to abstract mutable computations into a purely functional interface, harnessing the expressiveness and safety of the type system. The interface to manipulate the mutable ST boxes is very similar to what we have seen with TVars previously.

Rank-N Types

Again, this is a very advanced topic, and you do not need to understand them to be able to use the ST monad. Here is a very brief aside for the curious. Consider the following code:

```
-- A function that takes a polymorphic function and applies it
to a pair of values.
applyToBoth :: (forall a. a -> b) -> (b, b)
applyToBoth f = (f 1, f 'a')

-- A function f from above must work for ALL types uniformly;
it means
-- it cannot depend on the input, so can be a constant function
for instance
exampleF :: forall a. a -> Int
exampleF _ = 42

res :: (Int, Int)
res = applyToBoth exampleF
```

CHAPTER 12 DOWN THE RABBIT HOLE

The function f in the applyToBoth definition is a polymorphic rank-2 function with the (forall a. a -> b). This means that f must be able to take any type a as input and return a value of a specific b. Because f must work for any type a, it cannot make use of the specific contents of an input. Instead, it must treat every input uniformly.

There is not much practical use in the forall a. a type of definitions, aside from advanced tricks used in the ST monad. However, rank-2 (and rank-N) polymorphism can be quite useful in scenarios where you need to provide a high level of abstraction, particularly when working with higher-order functions or when you need more control over type inference and instantiation. Here are some real-world scenarios where rank-2 or rank-N polymorphism can be especially beneficial:

Callback functions: When you're writing higher-order functions that accept callback functions, you might use rank-2 polymorphism to ensure these callbacks can work generically over a range of types. This can be crucial in GUI frameworks or event-driven libraries where callbacks might need to handle different types of events:

```
registerCallback :: (forall a. Show a => a -> IO ())
 -> IO ()
registerCallback callback = do
    callback 42           -- An integer
    callback "Hello"      -- A string
    -- The callback function operates on any 'Show'able type.
```

The callback function that can be passed to registerCallback must work on every type that is a part of the Show typeclass.

Visitor pattern: The visitor pattern is a design pattern used in object-oriented programming where you might have a collection of elements, each of which can be visited and processed generically. In Haskell, you might leverage rank-2 polymorphism to ensure that the visitor function can handle any element type:

CHAPTER 12   DOWN THE RABBIT HOLE

```
data HTMLElement = forall a. Render a => HTMLElement a

renderPage :: [HTMLElement] -> IO ()
renderPage elements = mapM_ (\(HTMLElement elem) -> render elem) elements

class Render a where
  render :: a -> IO ()
```

Generic programming: Libraries like `lens` make heavy use of rank-N polymorphism to provide generic operations that work seamlessly over various data types. This leverages the ability to abstract over transformations that can apply generically, manipulating data structures in a highly reusable manner:

```
over :: ((forall f. Functor f => (a -> f b) -> s -> f t))
     -> (a -> b) -> s -> t
```

In each of these cases, the ability to express higher-level abstractions without committing to specific types enhances code flexibility and reusability. This often involves more complex type signatures but results in very powerful and abstract designs, making rank-N polymorphism a vital tool in an advanced Haskell developer's toolkit.

The main challenge of rank-N polymorphism is that Haskell cannot deduce the type signatures for you, so as soon as you start using it, you have to write out all types properly and carefully. But since it's a good practice anyway, this should not be such a big obstacle if you decide to put rank-N into your Haskell "spell book."

Vectors in Practice

But back to vectors! In practice, choosing between boxed and unboxed vectors, or deciding whether to use mutable vectors, largely depends on your specific use case. Do you need to store complex data structures with differing types? Boxed vectors might be your best bet. Or perhaps your task requires blazingly fast numerical computations with large datasets – unboxed vectors could be the way to go. For workloads demanding maximum speed, mutable vectors can deliver the performance needed by leveraging in-place updates.

The `vector` library, with its rich API and robust performance characteristics, exemplifies the power of Haskell's type system combined with GHC optimizations. By understanding these various aspects, you'll be well equipped to tackle performance-centric tasks efficiently and elegantly.

> **EXERCISES**
>
> 1. Refactor the mutable vector example to use unboxed vectors, measuring any performance improvements.
>
> 2. Design a simple matrix multiplication function using vectors and discuss boxed vs. unboxed implementations' trade-offs.
>
> 3. Implement a small simulation that uses mutable vectors to track and update a swarm of particles in 2D space.

More AI: Vector RAG

By now we have addressed almost all of the design goals for our AI framework: we redesigned the monad transformer stack in a very smart way to reuse all the previous code, we added MongoDB functionality, and we made sure the terminal and web UI can be used simultaneously. Sure, there are lots of feature improvements we can still add, but we have the foundation for us to develop the product in any way we want – all these new additions are just a technicality in the existing framework. We will discuss and implement some of them in the next, and last, chapter. For now, let's address the last goal – supporting so-called Vector RAG, or retrieval augmented generation.

As discussed in Chapter 9, LLMs perform much better when given proper context for interaction with the user. The easiest way to do it, while providing excellent quality,[2] is Vector RAG, described below (see also Figure 12-7). To give a simple example, if you are not Elon Musk or another very public persona, try asking any public chatbot "who is <my name>." You will get an "I don't know" answer or some sort of hallucination, depending on the settings. If on the other hand you provide some info about yourself, your project, your interests, etc. to the RAG-enabled AI agent, it will be able to respond the way you expect.

---

[2] There are more advanced ways to increase the quality of LLM output even more that are being actively researched. One of them is so-called Graph-RAG, where we store knowledge about the world in a graph database and then retrieve the relevant pieces together with unstructured Vector context. But this is out of scope for this book.

CHAPTER 12 DOWN THE RABBIT HOLE

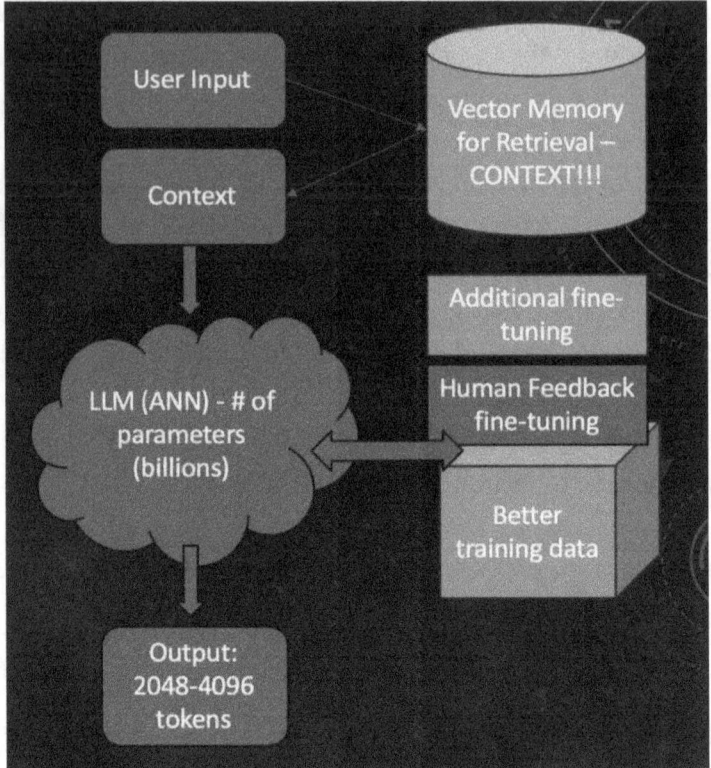

Figure 12-7. Vector RAG illustration

The way it works is via text-to-vector embeddings – a smart way to represent semantic information via multidimensional numeric vectors. Remember the queen = king - man + woman example from Chapter 9? Amazingly, the "sense" of words and whole concepts gets captured by this approach very well. So, to implement Vector RAG, we need to follow the following steps:

- Populate some database with relevant pieces of text, converted to vectors via embeddings API (we will use OpenAI for simplicity).

- When a user interacts with any of our AI agents, we convert their input into vector Vi as well and then perform vector similarity search.

- Compare Vi with all the vectors we have in our database by taking dot product (if our vectors are normalized to 1, and they are, it is equal to cosine of the angle between them, and so the closer the number to 1, the more similar are the vectors).

- Then pick one to five (or up to you) pieces of the most relevant text from our database and put them into the LLM's system prompt as part of the context.

This way, you can ensure that your agents "know" the information relevant to you and not just the vast corpus of publicly available data they were trained on. We will demonstrate the difference once we go through the current exercise.

Vector search is much better than simple keyword search, which was ubiquitous in the Google era – since it searches "by meaning," not by a number of keywords. There are now lots of different vector database vendors to support the increasing demand for Vector RAG; in fact, MongoDB provides this functionality as well. However, at the time of this writing, to get the vector search and indexing functionality in a local MongoDB, you have to deploy it as an Atlas cluster, which is a pretty advanced devops task. Plus, it is much more educational to build a simple vector search ourselves using Haskell – and then you can switch to using Mongo in production if you wish to do so. However, what we are going to build is more than enough for personal use.

CHAPTER 12   DOWN THE RABBIT HOLE

In-Memory Vector Engine

What we are going to build is very simple and straightforward In-Memory Vector Search Engine. Thanks to the design of the vector library, it will be very performant, and we will use MongoDB as persistent storage for the data. So the flow will work as shown in Figure 12-8.

Figure 12-8. *In-Memory Vector RAG*

WARM-UP EXERCISE

Experiment with different ways to implement the dot product function. Which ones perform better? Why?

Mongo Storage Part

Let's start with defining the data type that we will store both in Mongo and in memory for our vector embeddings:

```
-- src/Mongo/MongoRAG.hs
...
data RAGData = RAGData {
    title::Maybe Text,
    textContent :: Text,
    vectorContent :: U.Vector Float
} deriving (Show, Generic)
```

```
instance ToJSON RAGData
instance FromJSON RAGData

ragColName :: Collection
ragColName = "vectorRAG"
```

Everything is self-explanatory here – we have a data type that stores an optional title, text contents, and an embeddings vector. We derive the To/FromJSON instances as usual to be able to convert easier and define a Mongo collection name. Then, we need some low-level (IO) methods to insert new RAGData into Mongo and to find all existing RAGData (at startup):

```
-- LOWLEVEL interface
insertRAG :: MongoConnection -> RAGData -> IO Database.MongoDB.Value
insertRAG conn rgd = do
    let (Object jsn) = toJSON rgd
    let bsn = AB.bsonifyBound jsn
    withMongoConnection conn $ insert ragColName bsn

-- findAllSPD :: MongoConnection ->
findAllRAG :: MongoConnection -> IO [RAGData]
findAllRAG conn = withMongoConnection conn $ do
    cur <- find (select [] ragColName)
    bsn <- rest cur
    let results = map (fromJSON . Object . AB.aesonify) bsn
    let res = [rag | Success rag <- results]   -- Extract successful conversions
    pure res
```

The first method is a straightforward call to the Mongo interface to insert a RAGData object, converted first to JSON and then to BSON document. findAllRAG is a bit more complicated as we first get a Cursor

CHAPTER 12   DOWN THE RABBIT HOLE

and convert it to a list, then we have to map the backward conversion from BSON document through JSON to RAGData, but since each conversion is represented as Either data type, we have to do another filter run. This function can be optimized quite a bit – try that on your own.

Finally, we are using the type family trick we discussed previously in this chapter to create a uniform but typed interface to work with various Mongo collections:

```
-- src/Mongo/MidLayer.hs
{-# LANGUAGE TypeFamilies #-}
{-# LANGUAGE AllowAmbiguousTypes #-}
module Mongo.MidLayer
where

import Data.Kind (Type)
import Database.MongoDB (Value)
import Mongo.MongoRAG (RAGData, ragColName, insertRAG,
 findAllRAG)
import Data.Text (Text)
import MidMonad (Mid)
import Mongo.Core (MongoState(..))
import Control.Monad.MRWS
import StackTypes (Settings(..))

class MongoCollection a where
    type Record a :: Type
    colName :: Text
    insertOne :: Record a -> Mid Value
    findAll :: Mid [Record a]

instance MongoCollection RAGData where
    type Record RAGData = RAGData
    colName = ragColName
```

```
    insertOne rgo = do
        conn <- mainConnection <$> asks mongoSettings
        liftIO $ insertRAG conn rgo
    findAll = asks mongoSettings >>= (liftIO . findAllRAG) .
    mainConnection
```

Here we define an associated type family MongoCollection with two actions in our MRWST Mid monad – insertOne and findAll. Then we define an instance of this family for our RAGData type. Basically, all we have to do is extract mainConnection that stores our connection to Mongo from the Reader's settings and then run the low-level methods we defined above.

Finally, two methods in the src/Middleware.hs to use our newly built interface:

```
buildRAGM :: Mid (V.Vector RAGData)
buildRAGM = do
    -- Explicitly specify the type application for `RAGData`
    records <- findAll @RAGData
    return $ V.fromList records

-- openai --------------------------------
embedTextMid :: Text -> Mid ()
embedTextMid txt = do
    st <- get
    let prov = currentProvider st
    let mdlName = embeddingName (head embeddingModels)
    v <- lift $ embedText txt mdlName prov (loggerState st)
    _id <- insertOne @RAGData $ RAGData obj
    lgDbg $ "Successfully inserted " ++ show _id
```

The first one, buildRAGM, returns a boxed vector of RAGData objects that we will use as an in-memory store – so all we do here is call the method from MongoCollection type family and convert the result to Vector.

Note how we are calling the findAll method: `findAll @RAGData`. What the heck is this? Unfortunately, if we write simply findAll and hope for Haskell to figure out what the type of the result should be from the context, we will get a compiler error. So we have to tell `findAll` which type to set for the "a" type variable from the MongoCollection class – and this is done via @<TypeName> syntax. For this to work, you need to also enable the `{-# LANGUAGE TypeApplications #-}` extension.

> **Note** Using the TypeApplications extension, you can explicitly write which types to instantiate for polymorphic functions using func @<TypeName> syntax.

Even though it may seem there is enough information in the context to deduce the RAGData substitution, there are cases when it is not so, and in any cases writing `findAll @<MyDataType>` is cleaner and easier to maintain – then you know for sure what you will get as a result. It also looks cool.

The second function decodes the result of the OpenAI API embeddings call (the code is added to the LLM/OpenAI.hs file – there are no new concepts there, so feel free to peruse it on your own) and writes both the text and the resulting vector to the database. We have also expanded Commands.hs to include one more command – ":embed". You can now use it to embed any text to our vector storage. We will also expand `embedTextMid` later on to make sure it updates our in-memory storage, not just the database.

In-Memory Store Part

Now that we have prepared the MongodDB layer properly, let's make further adjustments to have the in-memory store up and running. First, let's adjust our global application state to include the in-memory store:

CHAPTER 12 DOWN THE RABBIT HOLE

```haskell
data AppState = AppState {
    httpManager :: Manager,
    currentModelId :: Text,
    currentProvider :: ProviderData,
    loggerState :: LoggerState,
    messageHistory :: [Message],
    historySize :: Int,
    uiState :: UIState,
    memoryStore :: MemoryStorage
}
```

Trivial (and we will look at the MemoryStorage type in a bit). Then, we very slightly adjust our initAll function to initialize the in-memory store from MongoDB – `memst <- findAllRAG $ mainConnection mongoState` – and add the resulting vector to the initial State variable (also trivial).

By now, everything is set up for us to be able to work with the in-memory store. First, let's define a function that will add a new item to the in-memory store – and we will call it after the insertion to Mongo. Look at the src/VectorStorage/InMemory.hs:

```haskell
{-# LANGUAGE OverloadedStrings, DuplicateRecordFields #-}
{-# LANGUAGE DeriveGeneric #-}

module VectorStorage.InMemory
where

import Data.Text
import qualified Data.Vector.Unboxed as U
import qualified Data.Vector as V
import Mongo.MongoRAG (RAGData (vectorContent))

type MemoryStorage = V.Vector RAGData
```

Chapter 12 Down the Rabbit Hole

```
-- Function to add a new RAGData item to the MemoryStorage
addRAGData :: MemoryStorage -> RAGData -> MemoryStorage
addRAGData = V.snoc
```

Luckily, the vector library provides a function to append an item to a vector, so we are simply defining a synonym. Of course this operation is not very efficient (O(n)), but since we will be inserting new items rarely, it doesn't matter; however, it makes a good exercise.

EXERCISE

Write a more efficient function for insertion data storage for RAGData. Hint: Use vectors that pre-allocate certain sizes and grow them in increments.

With this easy definition, all we need to do is adjust the embedTextMid function defined above to also update the in-memory store after inserting the new RAGData object into Mongo: `modify (\s -> s { memoryStore = addRAGData (memoryStore s) obj})`. That's it. Insertion of new items with embeddings defined via OpenAI API call is completely ready!

You can test the functionality by running Jarvis as usual (stack build, stack exec jarvis), typing or copying a couple of paragraphs of text, and running the :embed command, and you should see something like Figure 12-9.

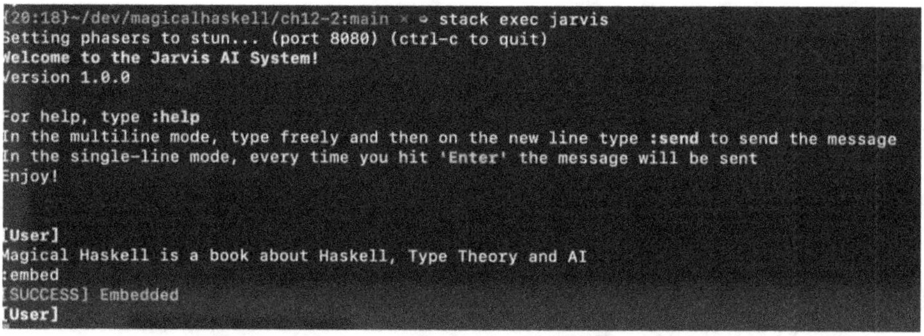

CHAPTER 12 DOWN THE RABBIT HOLE

```
{20:18}~/dev/magicalhaskell/ch12-2:main × ● stack exec jarvis
Setting phasers to stun... (port 8080) (ctrl-c to quit)
Welcome to the Jarvis AI System!
Version 1.0.0

For help, type :help
In the multiline mode, type freely and then on the new line type :send to send the message
In the single-line mode, every time you hit 'Enter' the message will be sent
Enjoy!


[User]
Magical Haskell is a book about Haskell, Type Theory and AI
:embed
[SUCCESS] Embedded
[User]
```

Figure 12-9. *Embedding new text in our vector storage*

To check that it's there, go to Mongo shell and run `db.vectorRAG.find()`.

Now we can turn to the exciting part – implementing vector search in our in-memory storage. Here is an outline of what we need to do:

- Convert what the user inputs to a vector by going to OpenAI embeddings API.

- Map over all our in-memory storage with the dot product function to calculate similarity.

- Sort the results in descending order.

- Take the top n results for further usage as context when interacting with AI agents.

EXERCISE

Try to implement the functionality above by yourself.

CHAPTER 12   DOWN THE RABBIT HOLE

Let's review the remaining low-level API code in the src/VectorStorage/InMemory.hs:

```
sortedSearchResults :: Int -> U.Vector Float
    -> MemoryStorage -> V.Vector (Text, Float)
sortedSearchResults n v ms =
    let sorted = sortedSearch v ms
    in  V.map (\(score, ind) -> (textContent (ms V.! ind),
    score)) (V.take n sorted)

sortedSearch :: U.Vector Float -> MemoryStorage -> V.Vector (Float, Int)
sortedSearch v ms = let x = dotProductAll v ms
                    in  sortVectorByFloat x

dotProduct :: (Num a, U.Unbox a) => U.Vector a
    -> U.Vector a -> a
dotProduct v1 v2 = U.sum $ U.zipWith (*) v1 v2

dotProductAll :: U.Vector Float -> MemoryStorage
    -> V.Vector (Float, Int)
dotProductAll v = V.imap (\i el -> (v `dotProduct` vectorContent el, i))

-- Function to sort a boxed vector of tuples (Float, Int) by the Float component
sortVectorByFloat :: V.Vector (Float, Int)
    -> V.Vector (Float, Int)
sortVectorByFloat = V.modify (sortBy (comparing (Down . fst)))

-- Function to sort a boxed vector of tuples (Float, Int) by the Float component
sortVectorByFloatM :: V.Vector (Float, Int)
    -> V.Vector (Float, Int)
```

CHAPTER 12 DOWN THE RABBIT HOLE

```
sortVectorByFloatM vec = runST $ do
    -- Create a mutable copy of the vector
    mvec <- V.thaw vec
    -- Sort the mutable vector in-place
    sortBy (comparing (Down . fst)) mvec
    -- Freeze the mutable vector back to an immutable one
    V.unsafeFreeze mvec
```

We have a bunch of interesting algorithmic code in this file. Let's review them in the logical order, bottom-up:

- dotProduct v1 v2 = U.sum $ U.zipWith (*) v1 v2: This is one of about ten possible implementations of the dot product for numerical vectors, but is the fastest. We are relying on tests we did a couple of years ago comparing performance of many of them against each other and C code here: https://medium.com/superstringtheory/once-more-on-haskell-vs-c-performance-a54498bfa91f.

- dotProductAll maps our dot product over the whole memory storage, and then there are a couple of sorting methods provided – one pure and one in the ST monad. Test them both to check the performance!

- Once everything is sorted, all we need to do is take the top n results and convert the intermediary (Float, Int) vector that stores similarity scores and indexes of the specific MemoryStorage element into actual (score, text) pairs using the sortedSearchResults function.

With this API in place – this is another highlight of Haskell, just several concise lines of code that implement quite advanced functionality – we can now implement an actual search function that will take the text from the current buffer, convert it to a vector, perform the search, and show the top results. To do this, we implement a searchRAGM function in the Mid monad:

```
-- src/Middleware.hs
...
searchRAGM :: Int -> Text -> Mid (V.Vector (Text, Float))
searchRAGM n txt = do
    st <- get
    let prov = currentProvider st
    let mdlName = embeddingName (head embeddingModels)
    v <- lift $ embedText txt mdlName prov (loggerState st)
    let ms = memoryStore st
    let res = sortedSearchResults n v ms
    pure res
```

This is straightforward access to various settings and state items and then running an embeddings API call and actual search/sorting from the low-level API. Then, we add a new command to the terminal UI:

```
-- src/Commands.hs:
...
commandSearchRAG :: [String] -> App()
commandSearchRAG (x:_) =
    let n = read x :: Int
    in _commandSearchRAG n
commandSearchRAG _ = _commandSearchRAG 5

_commandSearchRAG :: Int -> App()
_commandSearchRAG n = do
    txt <- currentLineBuffer <$> lift (gets uiState)
    if T.length txt > 0 then do
        res <- lift (searchRAGM n txt)
        uis <- lift (gets uiState)
```

CHAPTER 12 DOWN THE RABBIT HOLE

```
        lift $ modify (\s -> s { uiState = uis {
        currentLineBuffer = ""}})
        controlMessage [lgreen] "[Search Results:]"
        V.mapM_ (\(txt, score) -> controlMessage [white] ("["
        ++ show score ++ "] " ++ T.unpack (T.take 80 txt))) res
    else controlMessage [yellow] "[WARNING] Your message is
    empty, please type something first"
```

Again, just some housekeeping – getting a current text buffer, calling previously defined searchRAGM, and then showing the results along with the similarity scores.

Fire up Jarvis, add some RAG items via the :embed command, and then try searching for various queries to check the similarity (Figure 12-10).

```
[User]
what is Haskell?
:search
[Search Results:]
[0.5671976]
Magical Haskell is a book about Haskell, Type Theory and AI
[0.25925466]
hello world
```

Figure 12-10. *Testing Vector RAG search*

Isn't this cool? We built the Vector RAG from scratch in about 200 lines of code! The more relevant items you add (also optimal size should be several paragraphs of text), the more accurate the similarity score will become. Normally, with a relatively full database, you want to set the minimum score around .75 or even .8 for RAG – but it is better to always experiment with your own dataset to check the right parameters. In the screenshot above, there are only three very sparse items; hence, the scores don't look so good – but try copying some PDFs or just websites into your RAG and see what happens!

355

CHAPTER 12   DOWN THE RABBIT HOLE

Finishing RAG

With all the amazing goodies we have done in the previous sections, just searching is not enough – for RAG to be fully functioning, we need to send the context to our AI agent as part of the system prompt. This should be an easy exercise for you – just change the setup of [Messages] that are being already sent to OpenAI API with every :send call. Try to do it before reading further.

Here is one way to implement it, adding a convenience "context builder" function to Middleware.hs:

```
-- build context out of the RAG results with a minimum
threshold
buildContextM :: Float -> V.Vector (Text, Float) -> Mid Text
buildContextM minScore res = do
    let fres = V.filter (\(_, sc) -> sc >= minScore) res
    let txt = V.foldl' (\acc v -> acc `TX.append` "\n\n" `TX.append` (fst v)) "Please use the following additional context when answering the user but only if it is relevant:\n\n" fres
    pure txt
```

It takes the results of our vector search, filters them by the minimum similarity score, and then folds into a single text message.

Then in the actual :send command, we add a couple of lines to adjust our current [Messages]:

```
-- send the text that is currently in the text buffer
commandSendText :: [String] -> App()
commandSendText _ = do
    txt <- currentLineBuffer <$> lift (gets uiState)
    if T.length txt > 0 then do
```

```
-- getting RAG context and updating the Messages
   history:
ctx <- lift (searchRAGM 3 txt >>= buildContextM 0.6)
msgs <- lift (gets messageHistory)
lift $ modify (\s -> s {messageHistory = systemMessage
ctx : msgs} )
(asMsg, _) <- lift (chatCompletionMid $
userMessage txt)
uis <- lift (gets uiState)
lift $ modify (\s -> s { uiState = uis {
currentLineBuffer = ""}})
let fmt = highlightCode "Markdown" (T.pack asMsg)
outputStrLn (T.unpack fmt)
else controlMessage [yellow] "[WARNING] Your message is
empty, please type something first"
```

We run the RAG search, build context out of it, and then modify the messageHistory via usual helper functions in our Mid monad, nothing new here.

Now let's test it and compare results (Figure 12-11).

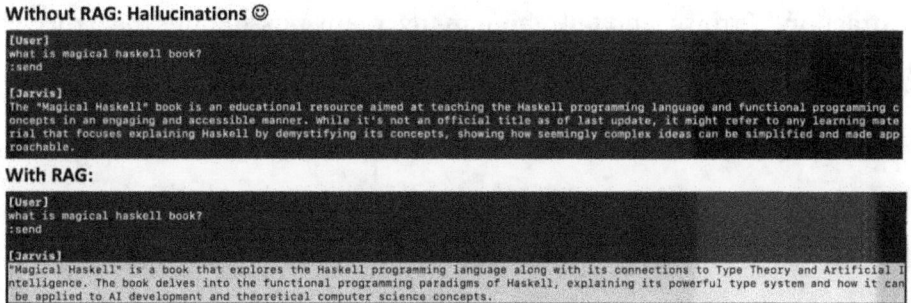

Figure 12-11. *Testing Jarvis with RAG about Magical Haskell*

CHAPTER 12 DOWN THE RABBIT HOLE

As you can see, without RAG, Jarvis hallucinates and invents some suggestions about what the *Magical Haskell* book might be. With RAG, where I entered some very basic info about this book, it responds nicely and to the point. Experiment with your own data!

EXERCISE

Add several new commands to the Jarvis terminal UI to handle various settings, e.g.: turn RAG on/off, change the minimum similarity score, change the number of found items to make a part of the context, etc.

Conclusion

This has been the longest chapter so far, but we learned and did so much – type families, advanced mutability, the ST monad, rank-N types, vectors – and we applied all that by building our own in-memory vector storage with dot product search support and implemented RAG-enabled AI agents on top of that. That is quite an achievement before the next, and last, chapter, where we will further enhance our AI framework; discuss the amazing "last abstraction," Arrows; and talk about many, many possibilities that you can make into reality using AI agents and the framework we have built!

CHAPTER 13

AI Multi-agents, Arrows, and the Future

What an amazing road have we traveled together, dear reader! We started from zero, pure mathematical nothingness. We learned how to construct increasingly complex types from basic numbers and symbols. We mastered the typed functional problem solving where we decompose the problem into smaller pieces and make sure the types match. We learned to build functions between these types. We introduced and built the typeclass hierarchy from the foundational Monoid to the Functor–Applicative–Monad triplet and to various useful extensions. We glanced over Type Theory and Category Theory and learned to represent our problem domain visually as a small category, with objects being types and typeclasses defining structure between those types. We "invented" the State monad, looked at its sisters – Reader and Writer – and built several increasingly complex monad transformer stacks to create end-to-end Haskell programs. We mastered the terminal UI and how to create web applications. We discovered mutability and designed our own amazing mutable RWST monad, which is usable for any web app out of box. We learned basic multi-threading and how to run both web and terminal UIs in one application. We delved into advanced mutable variables from

CHAPTER 13   AI MULTI-AGENTS, ARROWS, AND THE FUTURE

IORef, to MVar, to TVar, and to an amazing and a bit obscure ST monad. We built up our abstraction hierarchy to understand type families and learned how to use vectors efficiently. We even touched upon rank-N types and type application concepts. And the best part – we applied all of these very much advanced concepts in practice while building a practically useful AI framework that includes MongoDB, Vector RAG, and direct OpenAI access and fully built out a solid skeleton of a production app.

I hope you enjoyed this journey as much as I did – and even though Haskell is such a vast topic that we could have written ten times as many chapters and still not cover everything that it offers, my main goal was to try to teach you not to be afraid of the beautiful abstractions it is built upon and learn to love them. Once you do, you can catch up to any advanced topics on your own pace. I also hope that this book offered a good balance between theoretical considerations and applying them in practice so that you see their beauty and learn to use them when solving your programming problems.

However, there are still two topics left out of possible hundreds that I would like to cover in this last chapter:

- Arrows: A beautiful and very practical abstraction, underappreciated due to lack of accessible explanations and practical examples beyond parser libraries, FRP (functional reactive programming), and obscure DSLs. We will try to learn to not to be afraid of it in the first half of the chapter.

- Further practical extensions to our AI framework: We are still missing some pieces to make it truly production-ready, such as authentication for the web APIs, support of the advanced AI multi-agents as opposed to single-LLM "chatbots," error handling, and some others. We will try to cover what we can – and I hope that free tokens from integrail.ai will make it easier for you to experiment with what AI multi-agents can do.

So, dear reader, refresh your Haskell spell book, optionally review Chapter 3 as Arrows are closely connected to categories, and let us walk the last mile of our journey!

The code for this chapter is in the "ch13" folder of the book's GitHub repository.

Arrows: The Last Abstraction

If you are like me, you were probably curious about Arrows, even made several attempts to understand them, tried to read a couple of research papers discussing "not very relevant to practical day-to-day tasks" problems, got disappointed, and decided to stick to Monads. If that is the case, give yourself one more chance, and I will try to convey the beauty and simplicity of Arrows, taking into account everything we've learned so far. If you on the other hand have never heard about them, all the better – just remember everything we discussed when learning types, typeclasses, and categorical structure of our programs and let's try to enjoy the ride.

If you started thinking even a little bit in terms of our diagrams of composing, pun unintended, arrows, Haskell Arrows are the most basic of abstractions. In fact, if they were discovered a bit earlier, maybe we wouldn't even need monad transformer stacks and monads in general, as Arrows are *the last abstraction*, one ring to rule them all (only in a good sense).

CHAPTER 13   AI MULTI-AGENTS, ARROWS, AND THE FUTURE

Everything in typed functional programming, as we repeated in every chapter starting with the very first one, is to make sure our arrows compose (i.e., the types are right). Recall the diagram in Figure 13-1 about the Functor–Applicative–Monad triad.

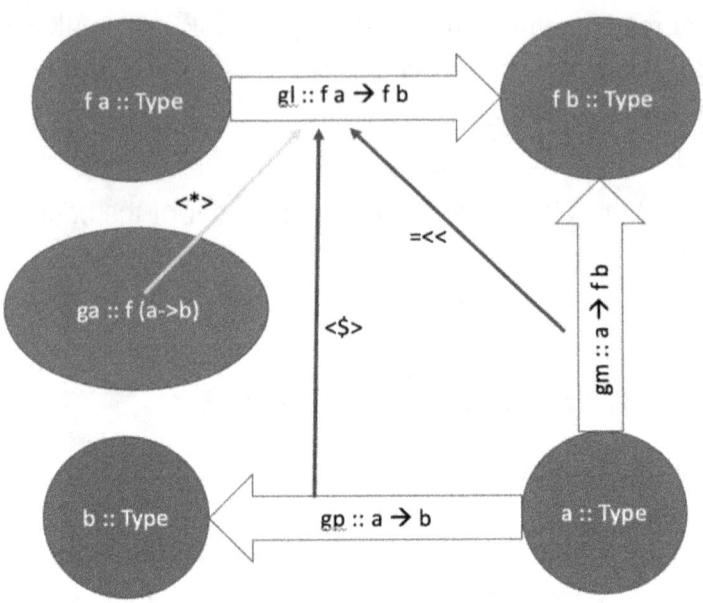

Figure 13-1. *How the Functor–Applicative–Monad hierarchy works*

We have pure types a and b and pure functions between them. We have "lifted" types f a and f b and even f (a -> b) – lifted type of pure functions (a -> b) – since functions are also values in Haskell, this is perfectly fine. Then we have functions between all these types. And then we have a beautiful symmetry, where main functorial (<$>), Applicative (<*>), and monadic (>>=) operators convert pure (a ->b), lifted (f(a->b)), and monadic (a -> f b) functions into the type of functions between lifted types (f a -> f b). All of it helps us make sure that every function (or arrow) composes with some other function (or arrow) while capturing various concepts and effects. All of the previous 12 chapters were centered around how to apply this diagram well.

Now, recall what a category is – it is simply a bunch of objects and arrows between them that compose! So, if we "zoom out" of our diagrams and forget what the internal structure of our objects (types) is and that our functions themselves can be of very different types, all that is left is just this: objects and composing arrows between them (Figure 13-2).

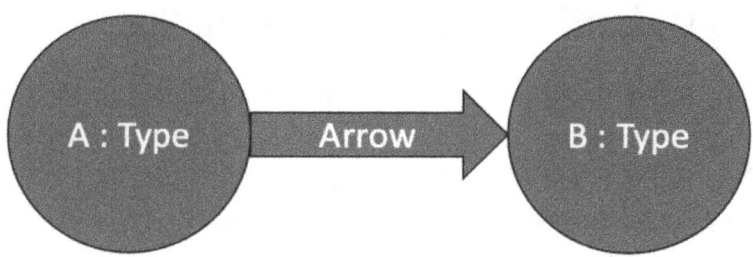

Figure 13-2. *An Arrow*

Hence, Arrows are the last, ultimate abstraction for typed functional programming. "But how are they different from a function?!", you may ask. Indeed, all functions *are* in fact Arrows! Just as well as all monadic actions – and we will show that in a minute. However, there are certain concepts that are impossible to design as monads, much less as pure functions – but they still can be represented as Arrows! Then, if we are able to define a nice abstraction, which captures pure functions, monadic actions, and even more complex arrows, but all using a single, unified interface, wouldn't it be wonderful?! Then we won't need to use one syntax for functions, another for monads with their do notation and what-not, and third for something else, but it will all be Arrows, nicely composing between types – just as our spells from Chapter 1.

Turns out, Haskell's Arrows are exactly this abstraction and interface. Another beautiful feature of Arrows is that they are extremely visual. However, this is also arguably their biggest drawback in Haskell language – writing what is inherently a 2D picture of types-objects and arrows between them as line-by-line and essentially 1D code can get quite confusing. I wish someone created a nice visual designer for Arrows interfaces – then people would be able to much easier appreciate their beauty and efficiency. Maybe it will be you, dear magician?

Arrows Definition and Basic Examples

But let's move step by step. Arrows are defined in Haskell via, very familiar to us by now, typeclasses. What can be made into an Arrow? Any data type (which we also called "type functions") that takes two other types as parameters, which are exactly the "a" and "b" objects (or types) between which our arrow lies. A prerequisite for a data type to be made into an Arrow is that it also has to be a Category typeclass, so we are really starting from the very basic foundations here:

```
class Category cat where
    id :: cat a a
    (.) :: cat b c -> cat a b -> cat a c
```

I am not including the Category laws here, but they are the same as for the mathematical definition of the category. As you can see from the definition above, "cat" can be any data type with two type variables, and in essence it defines an arrow between them. So `data MyCat a b = ...` can be made a Category if you can define the interface functions of course, and it describes an arrow between a and b. The Category typeclass defines two functions: an identity morphism (or, incidentally, arrow) `id` and composition operator (.) – which is basically the same as the function composition operator in Haskell that we used left and right previously. Since categorical (.) is a generalization of functional (.), this is totally fine. We will look at example definitions for pure functions and monadic actions soon, and it will become clear how these can be defined.

Now let's look at the Arrow typeclass, which adds additional useful interface functions:

CHAPTER 13 AI MULTI-AGENTS, ARROWS, AND THE FUTURE

```
class Category arrow => Arrow (arrow :: Type -> Type -> 
Type) where
-- lift a function into an Arrow
arr :: (b -> c) -> arrow b c
-- Send the first component of the input through the argument 
arrow, and copy the rest unchanged to the output.
first :: arrow b c -> arrow (b, d) (c, d)
-- Split the input between the two argument arrows and combine 
their output.
(***) :: arrow b c -> arrow b' c' -> arrow (b, b') (c, c')
...
```

There are many more useful operators in the Arrow typeclass, which we will review a bit further with diagrams that really ease the understanding. For now, let's look at how functions and monadic operators form a Category and Arrow so that you start building an intuition. We can do that since to fully define an Arrow typeclass only the arr and either first or *** from above are necessary, the rest are built from them (using the Arrow laws – so if and when you decide to build your own Arrows, be sure to review them in detail). Here is the definition of Category for the "normal functions":

```
instance Category (->) where
    id = GHC.Internal.Base.id
    (.) = (GHC.Internal.Base..)
```

As you would expect, for functions, which are just arrows (->), Category is trivial – id is id, and composition is composition. For Arrows:

```
instance Arrow (->) where
    arr f = f
    (f *** g) ~(x,y) = (f x, g y)
```

365

CHAPTER 13 AI MULTI-AGENTS, ARROWS, AND THE FUTURE

Again, trivial – "lifting" a pure function into a category of pure functions is just the function itself, and combining two functions into a function between pairs is just putting function applications inside a pair. No need to even draw anything.

With monadic actions, the situation is a bit more interesting. To turn any monad into Category/Arrow, we define an additional type:

```
newtype Kleisli m a b = Kleisli { runKleisli :: a -> m b }
```

The name Kleisli as you can guess comes from Category Theory research and may scare you off, but in essence we are just redefining any monadic action that has type `a -> m b` into a data type `(Kleisli m) a b` that we can make a Category and an Arrow, so here `cat = arrow = Kleisli m` but under the hood it's just still our good old monadic actions that we learned to use professionally. So how do we make them instances of Category and Arrow?

EXERCISE

Try to do it on your own before reading further: define Category and Arrow instances for Kleisli m.

We are skipping all of the other typeclass definitions for Kleisli and focusing on the topic at hand:

```
instance Monad m => Category (Kleisli m) where
    id = Kleisli return
    (Kleisli f) . (Kleisli g) = Kleisli (\b -> g b >>= f)
```

Again, if you look at it attentively and read the types, nothing scary: id simply lifts whatever is given into the monad via the monadic "return" operator, and composition is easy as well. We are pattern-matching the first monadic action to f and second to g on the left side and then pack a new monadic action as a result on the right. This new action takes an argument, applies g to it, unpacks it from the monad with the help of the >>= operator, and applies f to the result. Exactly the composition of two monadic actions. How about Arrow?

```
instance Monad m => Arrow (Kleisli m) where
    arr f = Kleisli (return . f)
    first (Kleisli f) =
         Kleisli (\ ~(b,d) -> f b >>= \c -> return (c,d))
```

The first one, `arr`, is to lift pure functions into our Arrow and here is again trivial – we are simply lifting the result of applying a pure function f into the monad, and this becomes an Arrow. The other, `first`, is a bit more complex, but as usual if you follow the types, it all becomes clear. It has to convert an Arrow between b and c into an Arrow between (b, d) and (c, d), where "d" is simply passed unchanged (this will become clear why it is useful in a bit). So what we do in the implementation is pattern-match our "hidden" monadic action in the first arrow to "f," apply it to "b," and return the results in a pair.

> **EXERCISE**
>
> Define an Arrow instance for Kleisli m via the *** operator instead of the "first" function as above. Remember, it has to combine two Arrows into an Arrow between pairs.

Okay, now that we have seen that both "normal functions" and monadic actions are indeed Arrows, let's review the full Arrows interface and learn to use it.

Arrows Interface

Let's review the main Arrows interface one by one.

Lift the pure function into an arrow (Figure 13-3):

```
arr :: (b -> c) -> arrow b c
```

Figure 13-3. *arr Arrow function, lifting a pure function into an Arrow*

What it means is easiest of all to imagine now using monadic actions – there we can also "lift" pure functions into monadic *context* and this way enrich our computations. As we shall see further, Arrows allow us to capture more sophisticated contexts for computations that are impossible to represent as a monad.

Send the first component of the input through the argument arrow, and copy the rest unchanged to the output (Figure 13-4):

```
first :: arrow b c -> arrow (b, d) (c, d)
```

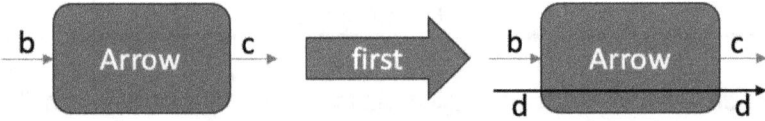

Figure 13-4. *first Arrow function, creating a new arrow that passes some argument untouched along with applying the original arrow*

I hope now it will start to become clear why visually representing the Arrows interface is very beneficial: from the above diagram you can easily see that we can "parallelize" computations, as we can plug in *two* different arrows into the output of `first someArr` while passing another argument unchanged, thus creating a "fork" of sorts – there are so many cases where it's needed as we shall see.

CHAPTER 13　AI MULTI-AGENTS, ARROWS, AND THE FUTURE

A mirror image of first, called unimaginatively second:

```
second :: arrow b c -> arrow (d, b) (d, c)
```

No need for a diagram. It's the same as above only the "d" arrow is on top of the b->c.

Split the input between the two argument arrows and combine their output (Figure 13-5):

```
(***) :: arrow b c -> arrow b' c' -> arrow (b, b') (c, c')
```

Figure 13-5. Arrow operator ***: combine two arrows into an arrow between pairs

Another ability to parallelize the computations, this time with applying arrows to both arguments (unlike first/second).

Fanout: send the input to both argument arrows and combine their output (Figure 13-6):

```
(&&&) :: arrow b c -> arrow b c' -> arrow b (c, c')
```

Figure 13-6. Arrow operator &&&: combine two arrows from one type into a "fork" arrow

Again, combining two arrows from one type to different types into one arrow from one type to a pair.

CHAPTER 13 AI MULTI-AGENTS, ARROWS, AND THE FUTURE

Then there are a bunch of composition operators as you would expect. Here we show two "forward" ones between arrows and between an arrow and a pure function, but there are also "reversed" variants for them as well (Figure 13-7):

```
-- composition of arrows, actually defined in the Category
typeclass
(>>>) :: Category cat => cat a b -> cat b c -> cat a c
```

Figure 13-7. Composition of two arrows

And with a pure function (Figure 13-8):

```
(>>^) :: Arrow arrow => arrow b c -> (c -> d) -> arrow b d
```

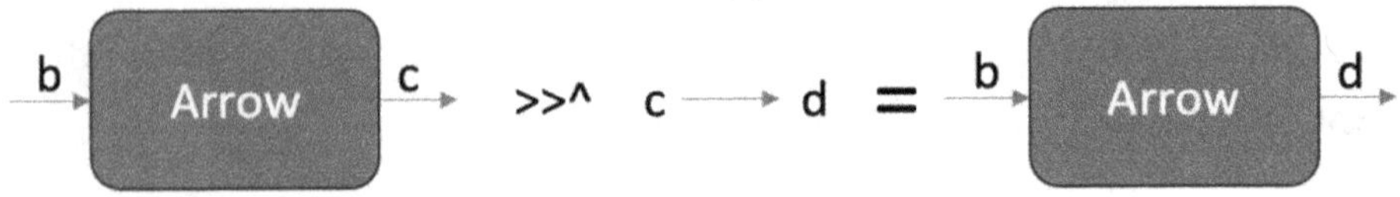

Figure 13-8. Composition of an arrow with a pure function

If you think about everything we learned about Haskell and this interface, represented visually, you may realize one simple thing – Arrows is the most natural way to represent Haskell (or even any typed functional for that matter?) programs! If you recall our "'M' Subchapter: Under the Hood of GHC" discussion, a Haskell program is a graph, which is being reduced during execution. Well, what are we creating using Arrows if not a graph of our program?! There is a beautiful 1:1 correspondence.

There are more additional typeclasses that provide lots of supporting functions, but here we will focus on the core interface to get familiar with Arrows – and then you can expand your spell book if you decide you like them.

CHAPTER 13 AI MULTI-AGENTS, ARROWS, AND THE FUTURE

Here it is important to realize one thing that people with imperative backgrounds stumble upon often. None of the combinators above actually *executes* anything. We are simply *composing* different arrows into one, huge Arrow-Graph that will represent our program. To "execute" it we need to give our Arrow some initial input and then have some sort of "execution" function, which will depend on the exact nature of our Arrows. For pure functions there's no need for anything additional; it's just function application. For monadic actions, recall the Kleisli definition:

```
newtype Kleisli m a b = Kleisli { runKleisli :: a -> m b }
```

This tells us that as soon as we built our program from Kleisli arrows by combining them into one superArrow, here is the way to run it: runKleisli superArrow initValue.

For other types of Arrows, it may be something else – we will see some examples below, but for now let us look at a simple pure-monadic program represented as Arrows to start appreciating the beauty on specific examples. Look at the src/Nodes/Simple.hs file in this chapter's source code directory (ch13):

```
module Nodes.Simple
where
import Control.Arrow

testMonad :: IO ()
testMonad = do
    putStrLn "Hi, what's your name?"
    nm <- getLine
    putStrLn $ "Hi, " ++ nm
```

CHAPTER 13 AI MULTI-AGENTS, ARROWS, AND THE FUTURE

First, let's look at the simple monadic program. We are outputting the prompt, asking the user for their name, and then printing a simple greeting. Everything is clear in the "imperative" do notation. Now let's look at the Arrow variant of the same:

```
testArrow :: Kleisli IO (String, String) ()
testArrow = first (Kleisli putStrLn >>> Kleisli (const getLine)) >>> arr (\(s1,s2) -> s2 ++ s1) >>> Kleisli putStrLn

runArrow = runKleisli testArrow
```

Or let's visualize this (Figure 13-9).

Figure 13-9. *Visualization of a simple Arrow-based test*

Let's uncouple this step by step. First, we are creating Arrows from monadic actions by simply wrapping them in the Kleisli constructor. Then, we create a composition out of putStrLn and getLine – which outputs something on the screen and then returns something that was input. Then we pass this composite Arrow to the `first` function, which splits it to be able to work with pairs; thus, our end result Arrow expects a pair (String, String) as input. The rest are a couple of simple compositions: we lift a pure function that combines two Strings into one with arr and then compose it with the final putStrLn Arrow. Study both the diagram and the code to fully understand it. The key here is that the whole "program" – `testArrow` – is completely point-free; it simply combines different Arrows together and then expects some input and produces some output. This may seem unfamiliar at first, but we will discuss a case next where such approach will be absolutely natural. To test this program, run stack ghci and

```
ghci> runArrow ("Hi, what's your name?", "Hi, ")
Hi, what's your name?
Anton
Hi, Anton
```

As you can see, works like a charm.

Of course for such a simple program, the whole shenanigans of converting everything into Arrows may not be worth it. You are free to read the papers and existing Arrows-based libraries for parsers and FRP on your own time – this can be very educational. In this book, we will consider a different real-world example where Arrows fit in naturally.

AI Multi-agents and Arrows

Time to get back to AI. What we've built so far in the previous chapters is a nice foundation for a potential agentic framework, but it still has lots of limitations. From the AI point of view, Jarvis is very basic: it works with just one OpenAI LLM. And even though we added the message history functionality for additional context as well as Vector RAG – two necessary improvements that already make Jarvis quite useful – proper AI multi-agents can do so much more.

CHAPTER 13   AI MULTI-AGENTS, ARROWS, AND THE FUTURE

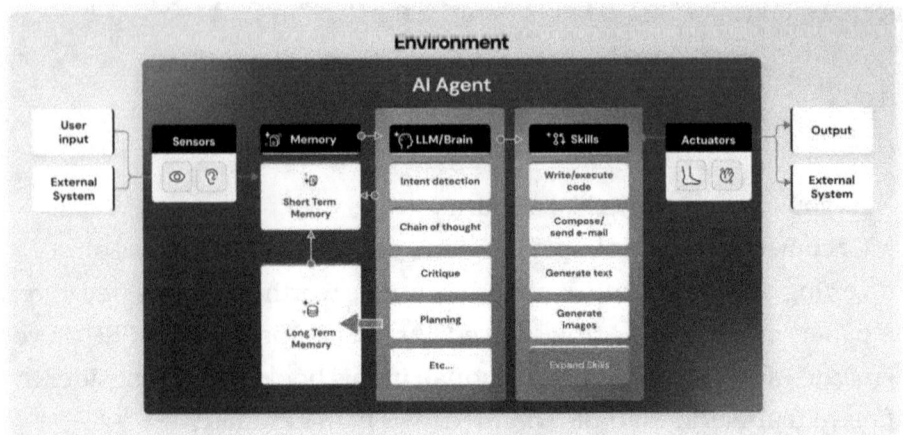

Figure 13-10. *Anatomy of the self-learning AI multi-agent*

Agents are like human beings in a sense. We have eyes (and other senses) to, uhm, sense the outside world. We have a mouth to communicate intelligently with each other. We have memory and (some) knowledge in our brain, and we also have some sort of reasoning engine, which makes decisions for us based on the processed senses and our knowledge. Last but not least, we have hands (and other body parts) to enact some sort of change in the external world.

But that's exactly what we want in the AI agent as well! Can ChatGPT do all of the above? Of course not, it can only pretend to "talk" to you. What we want in an agent is what is shown in Figure 13-10. Let's review this "anatomy" part by part. This is a very general scheme, but it captures the high-level design of pretty much any AI agent we can imagine – self-driving car, autonomous robot, agent on the Web, etc. Let's limit ourselves to the agents we can create on the Internet and software and ignore robots for now. It shall have/should be able to do the following:

Sensors: To sense the environment our agent operates in, it needs to be able to

CHAPTER 13   AI MULTI-AGENTS, ARROWS, AND THE FUTURE

- Read the websites, preferably the way humans do (as the websites have been built for humans, not for robots).

- Discover and be able to use various APIs available on the Internet and designed for computers.

To build such sensors, we will need a combination of LLMs (large language models) and models that can understand images (or convert them to textual descriptions) plus a little bit of software code that can crawl and download data at different URLs. Then with the right prompt engineering, our LLM-based sensors will convert what they "read" into formats suitable for further processing. And Haskell is very good at data processing.

Human interaction: This is the part everyone is familiar with by now thanks to ChatGPT – you can type what you want or say it with words, and it will be processed by another LLM-based module for further reasoning.

Process input: This, also LLM-based, module takes whatever sensors give it together with the current request from a human and tries to formulate a clear task request for our plan solution module – arguably, the main part of the "brain." It is also absolutely critical for this module to consult the knowledge base/memory via RAG and make it part of the context when formulating the task request.

Knowledge base/RAG: This module stores the data and knowledge that may be relevant to our agent's operations. This can be all kinds of publicly available data accessed via "regular" Internet search, as well as so-called vector databases, which represent unstructured text via numerical vector embeddings. This provides a much faster search "by meaning" as opposed to simply "by keywords" and is a crucial part of our agent. We have built a simple version of this in the previous chapter.

Plan solution: This is normally a bunch of different LLMs working together, as it has to take the task request, analyze what kind of resources (and "hands") are available to the agent to execute this request, iteratively

CHAPTER 13   AI MULTI-AGENTS, ARROWS, AND THE FUTURE

plan such an execution using various critique/step-by-step planning approaches, design sub-agents that are currently missing but needed for the task execution, and finally orchestrate execution using the "hands" or sub-agents available to our agent. This is an extremely interesting and fast-developing area of AI research. In some other, more specialized agents (e.g., in games), approaches such as reinforcement learning are quite useful as well.

"Hands" or execute actions: All of the above would be completely useless if our agent didn't have "hands" to do something that a human asked of it. These hands are pieces of code that can call external APIs, click buttons on web pages, or in some other way interact with existing software infrastructure.

I am sure that by now, with the foundation we have developed in code and the Haskell spell book we learned to use in the previous chapters, you are able to expand our AI framework further to create any kind of agents you like. We will cover a couple of additional practical pointers in the second half of this chapter that may help you on this journey.

For now, I would like to take a shortcut and introduce the **integrail.ai** platform to you – we are building it specifically to make creation of new AI agents easy, fast, and fully visual. All agents you may create there have an automatic API, so you can integrate them into the AI framework we built and make it much more powerful. Oh and by the way, if you use the code "magicalhaskell" when registering at Integrail, we will give you $20 worth of credits for your GenAI/Agentic experiments – this is more than enough to create and fine-tune several very useful agents for your needs.

However, my fully disclosed interest to promote our platform is not the only and even not the main reason for introducing it here. As I mentioned, it turns out that AI multi-agents architecture fits into the Arrows interface perfectly. Let me demonstrate how and let's build a simplified, but quite serious, Arrow framework from scratch to support it.

CHAPTER 13   AI MULTI-AGENTS, ARROWS, AND THE FUTURE

AI Multi-agents Examples

Let's look at a couple of examples of real-world multi-agents (Figures 13-11 and 13-12).

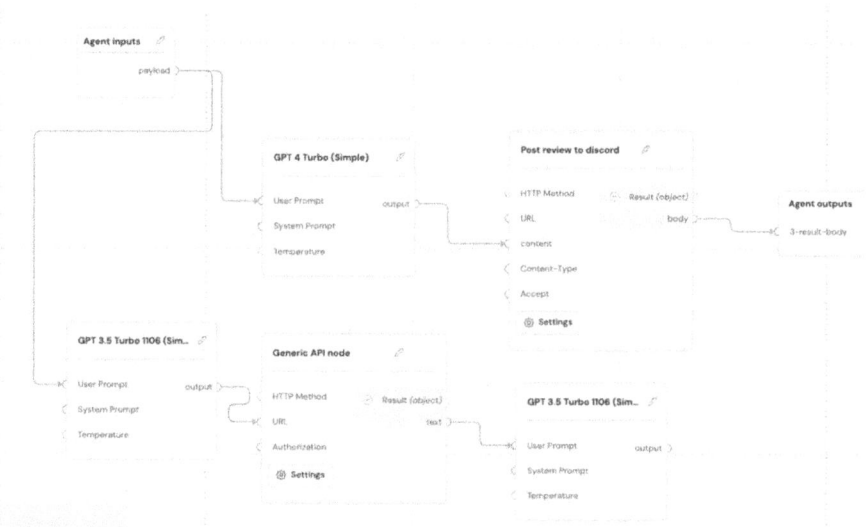

Figure 13-11. *AI multi-agent that reviews code pull requests on GitHub*

The above is an example of a real-world agent we use in production to review the pull requests our developers make. Every agent starts with "Agent inputs" and finishes with "Agent outputs." You can examine them directly or access via API. The other boxes on the diagram we call "nodes," and they perform a specific action, but also with very clear input and output format. This agent integrates with GitHub, analyzes the PR via LLM nodes, provides comments, and puts them directly into GitHub without any human interaction.

By connecting the nodes while respecting the types, we get an "execution pipeline," which we can then execute either by clicking the "play" button in the interface, by using chat UI, or by making an API call.

CHAPTER 13   AI MULTI-AGENTS, ARROWS, AND THE FUTURE

Doesn't it remind you of something? Nodes look just like Arrows – input, output, some execution context. Connections between them are arrow compositions. Let's look at a more generic AI agent that does not integrate with external systems that much, but mimics ChatGPT functionality with some additional quirks.

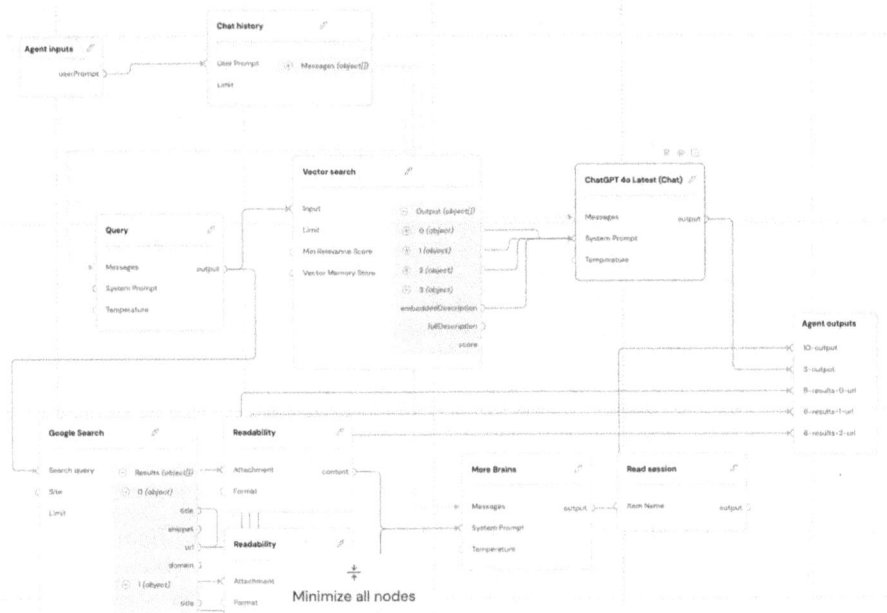

Figure 13-12. AI chat agent with history, Internet search, and Vector RAG

Let me explain what's going on here so that you may start to appreciate the agent design methodology that you will need to master regardless of whether you use Haskell or visual tools such as Integrail to build them.

The user's textual input is being combined with the Chat history node to build part of the context (we have this implemented already in our framework). Then it goes to a standalone LLM node called "Query," whose goal is to build a good search query from the user's request. We are currently skipping this step in our framework, and it is much more critical

CHAPTER 13   AI MULTI-AGENTS, ARROWS, AND THE FUTURE

for Google search vs. our own vector search, but it's still a best practice. The query goes in parallel to two other nodes – Vector search and Google Search, which do what you would expect. Vector search returns context directly (just like in our current Haskell code), while Google Search needs other nodes – called "Readability" – to read and convert the URLs that were found to markdown format, which is easier for LLMs to process. Finally, all this additional context is given to the "More Brains" LLM node, which processes the user's request with excellent precision, since the context is relevant.

Building agents of pretty much arbitrary complexity takes very little time on Integrail – you are simply dragging and dropping the nodes, connecting them respecting the types, and editing the prompts for LLM nodes so that they do what you ask of them. I will explain how to integrate with your Integrail agents using Haskell in the second part of this chapter, but for now let's go back to Arrows.

Arrows to Support Quick AI Multi-agent Creation

Okay, we can build AI agents visually, but what if we want to almost just as quickly build them in Haskell to then integrate with my local PC context and so on? Again, the similarity of this visual approach to design agents and the Arrows interface is striking:

- **Nodes are Arrows**. They take an input of type a and return an output of type b, where types are of course different for different kinds of nodes: LLMs, text-to-image/image-to-text models, text to voice/voice to text, web search, RAG search, document conversion, external API integrations, etc. All of them do different things, but all of them are just that – Arrows!

CHAPTER 13   AI MULTI-AGENTS, ARROWS, AND THE FUTURE

- **Connections between nodes are Arrow compositions.** We connect different nodes between each other in the order we need, respecting the types, and we get a final "composite" Arrow – an agent itself that has only one input and one output type!

This tells us that Arrows might be an excellent approach to create distributed execution pipelines that include not just external API calls, but also "internal" Haskell code, OS commands, etc. – in a uniform fashion. In the next section, we will sketch out a framework that can serve an excellent starting point for your further experiments whether you want to build connections to various LLMs and other GenAI models yourself completely in Haskell code or you are willing to take shortcuts and use external AI agents as part of your library of external APIs. Our goal is to make it absolutely universal and vendor independent.

So what is such a "node" is in an AI multi-agent workflow? It is just some code that takes the input in the defined format and outputs some other type of data, so, again, an Arrow. However, a question comes up – why can't we simply write monadic code, get "free" Arrows via Kleisli, and be done with it?

Well, it turns out, we want to represent *more* context than is possible using monads when we are designing such an approach to agents!

Arrows can represent more context than is possible with monads. Examples of that are in the various Arrows papers, such as the original paper from Hughes, as well as in the case at hand: AI multi-agents.

Let me give a simple example. Let's say I have a simple LLM-based agent that uses RAG for context. If RAG search doesn't find anything relevant, I want my agent to use some default context. How do I set this default? Of course I can write a monadic function that does RAG search

CHAPTER 13   AI MULTI-AGENTS, ARROWS, AND THE FUTURE

(just as we do now) and that checks if the output is empty and then puts some default context in that case – but, first of all, this seems clumsy and, second, what if I want to provide *different* default outputs to *the same type* of node in different agents? Do I write a separate monadic action for every agent? Do I pass the defaults as part of monadic state? It is terrible – if I have many nodes, I'll have to create just as many default fields in the state or resort to similar ugly solutions.

Turns out, Arrows provide a much more elegant way. Take a look at the src/Nodes/Arrows.hs file:

```
{-# LANGUAGE RankNTypes #-}
{-# LANGUAGE InstanceSigs #-}
{-# LANGUAGE GADTs #-}

module Nodes.Arrows
where

import Control.Category ( Category(..) )
import Control.Arrow
import Control.Monad ( (>=>), liftM2 )

data Node m e a b = Node {
    runNode  :: a -> m (Either e b),
    defaults :: m (Maybe b)
}
```

We are interested in the Node m e a b data type, which will represent our nodes and which we will make into Arrows. Of course, the Arrow will be defined for Node m e, since an Arrow is defined always between two types. Here "m" is a monad, whose monadic actions we will use under the hood to provide functions such as external API calls, logging, etc., while "e" is a type of errors we shall use. This is critical, as running nodes in AI multi-agents may sometimes result in errors, since we are relying on external APIs quite a bit – and we need to handle them gracefully.

The Node-Arrow data type itself contains two fields:

- runNode :: a -> m (Either e b): This is the main "node-running" function that will talk to an LLM, do Google search, read Gmail, post to GitHub, etc. – whatever nodes we want to provide for our agent. It takes an input of type "a" and returns a monadic value of Either e b. If you remember what Either is, it provides two constructors, Left and Right, and people use it as one approach to handle errors. If it returns Left e, it's an error; if it returns Right b, it's our proper output value.

- defaults :: m (Maybe b): This is where we will be able to assign default outputs for our node, *completely independent* of the actual code of the node! This gives us flexibility inaccessible with monads. In case there are defaults, we put them as Just def; if there are none, we put Nothing. This way, we will be able to build flexible agents, which provide default outputs in case of a failure – just as a simplified case described above.

Now that we have defined our data type, let's make it an instance of Category and Arrow. We purposely do not optimize the code here so that it's easier to comprehend what's going on, but once you understand, please do get rid of all these nested "case" statements as we did many times previously:

```
instance (Monad m, Monoid e) => Category (Node m e) where
    id :: Node m e a a
    id = Node (pure Prelude.. Right) (pure Nothing)

    (.) :: Node m e b c -> Node m e a b -> Node m e a c
    Node fbc dc . Node fab db = Node {
        defaults = dc,
        -- executing first function first
```

CHAPTER 13 AI MULTI-AGENTS, ARROWS, AND THE FUTURE

```
    runNode = \xa -> fab xa >>= \rb ->
        case rb of -- analyzing its result
            -- if there's error then:
            Left err -> db >>= \db' ->
                case db' of
                    -- no defaults? Just pass the error
                    Nothing -> pure (Left err)
                    -- defaults are present? apply next
                       function to them
                    (Just xb) -> fbc xb
            -- if there's no error just apply the next one
               to the results
            Right xb -> fbc xb
}
```

The id function is straightforward: we translate the argument without any changes in the runNode field through (pure . Right) and we set (pure Nothing) as defaults.

The composition function (and (.) is equal to the (>>>) operator) is slightly more complicated. For defaults of composing two arrows, we are simply putting the defaults of the last applied arrow – this makes sense if you think about it. For runNode, please carefully read the code above with the comments and you will understand the flow. Basically the main idea is in the error handling and how/when we look at the defaults. This is important – as it demonstrates this ability to manipulate additional context, which is completely impossible in the monadic interface. Dwell on it until you understand why. Monoid e restriction is currently not used, but we may want to make errors a Monoid to "stack" them similar to the Writer monad as we go along.

Let's now switch to the Arrow implementation. Here, we are not just defining the mandatory first and arr, but also the other combinators, as we want to make sure the defaults are handled the way we want:

```
instance (Monad m, Monoid e) => Arrow (Node m e) where
    arr :: (b -> c) -> Node m e b c
    arr f = Node {
        runNode = \x -> pure (Right $ f x),
        defaults = pure Nothing
    }
```

Lifting a pure function is easy – no defaults, put the result of its application under the Right constructor inside the monad. For the rest of the combinators, please read the comments carefully as they describe the logic:

```
first :: Node m e b c -> Node m e (b, d) (c, d)
    first (Node fbc dc) = Node {
        runNode = \(xb, xd) -> fbc xb >>= \xc -> -- apply
        function first
            case xc of
                -- if error:
                Left err -> dc >>= \dc'->
                    case dc' of
                        -- if no defaults - translate error
                        Nothing -> pure (Left err)
                        -- otherwise - translate defaults
                        (Just dcc) -> pure (Right (dcc, xd))
                -- if no error: return the result
                Right xcc -> pure (Right (xcc, xd)),
    -- this might be questionable, but we have no way
       of knowing
    -- which defaults to give to the "d" parameter, so
```

CHAPTER 13 AI MULTI-AGENTS, ARROWS, AND THE FUTURE

```
    -- even if we have defaults for "c", we assume there's
       no defaults
    -- going forward.
    -- have to test on different pipeline configurations
    defaults = pure Nothing
}

-- Split the input between the two argument arrows and
   combine their output.
-- redefining because we need to treat the defaults
   correctly!
(***) :: Node m e b c -> Node m e b' c' -> Node m e (b, b')
(c, c')
(Node fbc dc) *** (Node fb'c' dc') = Node {
    -- if we have both defaults, combine; if not - Nothing
    defaults = dc >>= \dcm -> liftM2 (,) dcm <$> dc',
    runNode = \(xb, xb') -> do
        resc  <- fbc xb
        resc' <- fb'c' xb'
        case resc of
            Right rc ->
                case resc' of
                    Right rc' -> pure $ Right (rc, rc')
                    Left err  -> pure $ Left err
            Left err -> pure $ Left err
}

-- Fanout: send the input to both argument arrows and
   combine their output.
-- redefining because we need to treat the defaults
   correctly!
```

CHAPTER 13   AI MULTI-AGENTS, ARROWS, AND THE FUTURE

```
(&&&) :: Node m e b c -> Node m e b c' -> Node m e
b (c, c')
(Node fbc dc) &&& (Node fbc' dc') = Node {
    -- same as *** !
    defaults = dc >>= \dcm -> liftM2 (,) dcm <$> dc',
    runNode = \xb -> do
        resc  <- fbc xb
        resc' <- fbc' xb
        case resc of
            Right rc ->
                case resc' of
                    Right rc' -> pure $ Right (rc, rc')
                    Left err  -> pure $ Left err
            Left err -> pure $ Left err
}
```

The `first` operator for Node operates on a type b to produce a type c (i.e., Node m e b c) and transforms it into a Node that operates on a pair (b, d) to produce a pair (c, d):

- The function runNode for "first" applies the original Node function fbc to the first element of the input pair (xb, xd).

- If the result "xc" is "Right", it means the computation was successful, and the output is a pair "(xcc, xd)", preserving the second element of the input tuple.

- If "xc" is "Left" (i.e., an error), the operator checks for a default value "dc'" using the defaults function "dc".

- If defaults are present ("Just dcc"), it returns the output as a pair "(dcc, xd)".

- If no defaults are present ("Nothing"), the error is propagated as "Left err".

CHAPTER 13 AI MULTI-AGENTS, ARROWS, AND THE FUTURE

The defaults function always returns Nothing. This reflects a limitation where the component associated with d doesn't have a defined default.

The *** operator for the Node type constructs a new Node that takes two Nodes and combines them to operate on pairs of inputs (b, b') and produce pairs of outputs (c, c'):

- Apply their respective "runNode" functions "fbc" and "fb'c'" separately on components "xb" and "xb'" of the input pair "(xb, xb')".

- If both computations succeed (i.e., results are "Right rc" and "Right rc'"), combine them into a

  "Right (rc, rc')".

- If either computation results in an error (e.g., one returns "Left err"), the final result is "Left err". It suggests that the propagation of errors happens on an "either-or" basis: any error in the chain leads to an overall failure.

For defaults, the code uses liftM2 (,) to attempt combining both Nodes' defaults into a single pair (dcm, dc'm). If both nodes provide defaults, a tuple of defaults is returned; otherwise, Nothing. This is a nuanced choice and assumes proper default handling when both components have defaults available.

By defining these operators, the code sets up a framework for combining and sequencing computations (with monadic side effects thanks to "m") in a modular and error-aware way, considering both the computation results and optional default values.

We will not be developing this code further in the book due to format limitations, but by now you should both have seen how to define your own Arrow types and appreciate the fact that you can operate pretty much *any*

CHAPTER 13   AI MULTI-AGENTS, ARROWS, AND THE FUTURE

additional context inside Arrows, unlike the monads. However, we do plan to develop an Arrow-based Haskell AI open source library at Integrail and make it available – so stay tuned for the updates.

For now, here are some examples of how an AI multi-agent pipeline can be defined using the Arrows interface and the Node type similar to that defined above. Let's say we want an agent that takes a user input including text and image, converts the image to text, does vector RAG, builds complex context, and uses an LLM to respond. Then, assuming we have actual monadic code for running different nodes, we could write something akin to (pseudo-code, don't try to compile it, just to give an example):

```
-- Assume some monad over IO - App and String for errors type
-- Extract text from user input
userInputToText :: UI -> Text

-- Extract image url from user input
userInputToImage :: UI -> ImageURL

-- create an arrow that extracts both from the user input
stage1 :: Node App String UI (Text, ImageURL)
stage1 = (arr userInputToText) &&& (arr userInputToImage)

-- Node / Arrow that searches for context - with defaults!
-- Here we have created a node with reusable code, but custom
defaults, showcasing Arrows power
vectorRAG :: Node App String Text Text
vectorRAG = Node {
      runNode = <search for context code as usual>
      defaults = pure $ Just "Some default context"
}
```

CHAPTER 13 AI MULTI-AGENTS, ARROWS, AND THE FUTURE

```
-- Another arrow with defaults that converts the image to text
-- if there is no image or we couldn't decipher - also
providing defaults so that agent execution doesn't stop here!
imageToText :: Node App String ImageURL Text
imageToText = Node {
      runNode = <image to text API call code as usual>
      defaults = pure $ Just "There was no image provided"
}

-- combining previous two arrows to be able to plug them
into stage1:
stage2 :: Node App String (Text, ImageURL) (Text, Text)
stage2 = vectorRAG *** imageToText

-- now we can do - "context prep stage":
ctxPrep = stage1 >>> stage2
-- they will compose, check the types!
-- or, avoiding middle steps shown here for simplicity:
ctxPrep = (arr userInputToText) &&& (arr userInputToImage)
         >>> vectorRAG *** imageToText
-- then we need an arrow to combine context, plug it into the
LLM Node / Arrow - and we are done!
```

I hope the example above is able to whet your appetite about what Arrows can do. It is such a beautiful way to define complex execution pipelines. Of course, the framework presented above is still very simplistic – it needs better error handling, and it needs more defaults – not just for outputs in case of errors, but also for inputs in case they are not provided by previous nodes and so on. Implementing these should be an enjoyable exercise in applying everything you learned about Haskell so far – or join one of our open source projects and let's work on it together!

CHAPTER 13   AI MULTI-AGENTS, ARROWS, AND THE FUTURE

Final Improvements for the AI Framework

Before we conclude this book with a recap of the functional typed program design, let's quickly address some pointers toward how the Jarvis AI framework can be further improved.

Web Authentication with JWTs

We have only started developing the web API in this book, but the code base and the discussion from Chapter 11 should give you more than enough to improve and extend it any way you like. The thing that's missing is authentication. There are many different ways to approach it, but using so-called JSON Web Tokens (JWTs) is arguably the most popular and one of the easiest. To implement JWT authentication in our framework, you need to issue tokens to your API users and then check their validity with methods similar to what we provided in src/WebAPI/JWT.hs:

```
{-# LANGUAGE DeriveGeneric, ScopedTypeVariables #-}
{-# LANGUAGE OverloadedStrings, DuplicateRecordFields #-}

module WebAPI.JWT
where

import GHC.Generics (Generic)
import Data.Aeson (ToJSON, FromJSON, Object, genericToJSON,
defaultOptions, genericParseJSON, toJSON, Value(..))
import qualified Data.Aeson as Aeson
import Data.Text (Text, unpack)
import Network.Wai (Request, requestMethod, rawPathInfo,
rawQueryString, requestHeaders, requestBody)

import qualified Data.ByteString.Lazy as BL
import qualified Data.Text as T
import qualified Data.Text.Lazy as TL
```

```haskell
import Data.Text.Lazy.Encoding (decodeUtf8)
import qualified Data.Map as Map
import Control.Monad.IO.Class (liftIO)

--import Web.Scotty
import Web.Scotty.Trans
import qualified Web.JWT as JWT

import Network.HTTP.Types.Status
import MidMonad (Mid)
import Control.Monad.MRWS (lift)
import Middleware (lgDbg)

-- change this in production!!!
mainSecretKey :: Text = "jarvis-new-generation-secret-token"

-- Define the Node data type
data UserToken = UserToken
  { user       :: Text
  , token      :: Text
  } deriving (Show, Generic)

-- Define the main data type
data TokenResponse = TokenResponse
  { status :: Text
  , result   :: UserToken
  } deriving (Show, Generic)

-- Derive ToJSON and FromJSON instances
instance ToJSON UserToken
instance FromJSON UserToken

instance ToJSON TokenResponse
instance FromJSON TokenResponse
```

CHAPTER 13 AI MULTI-AGENTS, ARROWS, AND THE FUTURE

```haskell
appUserToken :: ActionT Mid ()
appUserToken = do
            req <- request
            let maybeToken = extractBearerToken req
            case maybeToken of
                Nothing    -> liftIO $ putStrLn $ "No valid
                Bearer token found"
                Just token -> do
                    -- lift $ logDbg $ "Token: " <> (Data.Text.
                       unpack token)
                    let dtk = JWT.decodeAndVerifySignature
                    (JWT.toVerify . JWT.hmacSecret $
                    mainSecretKey) token
                    case dtk of
                        Nothing -> liftIO $ putStrLn $
                        "Illegal token!"
                        Just goodJWT -> do
                            let cl = JWT.unClaimsMap $ JWT.
                            unregisteredClaims $ JWT.
                            claims goodJWT
                            -- lift $ logDbg $ show goodJWT
                            lift $ lgDbg $ "Claims: \n"
                            ++ show cl
                            -- ok creating a user token
                               to return
                            let cs = mempty { -- mempty returns
                            a default JWTClaimsSet
                                JWT.iss = JWT.stringOrURI .
                                T.pack $ "jarvis-haskell"
```

CHAPTER 13   AI MULTI-AGENTS, ARROWS, AND THE FUTURE

```
                        , JWT.unregisteredClaims = JWT.
                        ClaimsMap $ Map.union (Map.
                        fromList
                            [("user", String
                            "email")]) cl
                        }
                    let key = JWT.hmacSecret
                    mainSecretKey
                    let newToken = JWT.encodeSigned key
                    mempty cs
                    -- lift $ logDbg $ show newToken
                    lift $ lgDbg "Returning new token"
                    json $ TokenResponse {
                        status = "ok",
                        result = UserToken
                            { user = "user"
                            , token = newToken
                            }
                    }

-- Function to extract the bearer token from the
Authorization header
extractBearerToken :: Request -> Maybe Text
extractBearerToken req = do
    authHeader <- lookup "Authorization" (requestHeaders req)
    let token = decodeUtf8 (BL.fromStrict authHeader)
    if "Bearer " `TL.isPrefixOf` token
        then Just (TL.toStrict (TL.drop 7 token)) -- Drop
        "Bearer " prefix and convert to strict Text
        else Nothing
```

CHAPTER 13 AI MULTI-AGENTS, ARROWS, AND THE FUTURE

This appUserToken is an action within Scotty's ActionT on top of our Mid monad, which deals with HTTP requests. Here's what it does:

- Request handling: The function begins by obtaining the current HTTP request using "req <- request".
- Token extraction: It tries to extract the "Bearer" token from the "Authorization" header of the request using the "extractBearerToken" function. The result is stored in "maybeToken".
- Token validation:
  - If "maybeToken" is "Nothing", it logs an error message "No valid Bearer token found."
  - If there is a token ("Just token"), it attempts to decode and verify the JWT using a HMAC secret key ("mainSecretKey"). This is done with "JWT.decodeAndVerifySignature". The result is stored in "dtk".
  - If "dtk" is "Nothing", it means the token is invalid or cannot be verified, and it logs "Illegal token!"
  - If "dtk" is "Just goodJWT", the token is valid. It then extracts the claims from the token and logs them.
- Token generation:
  - It starts preparing a new JWT with a specific claims set defined by "cs", which includes "iss" (issuer) as "jarvis-haskell" – change it to yours! It also adds a "user" claim, which is set as "email".
  - It uses "JWT.hmacSecret mainSecretKey" again to sign a new JWT.

CHAPTER 13   AI MULTI-AGENTS, ARROWS, AND THE FUTURE

- It encodes the new token with "JWT.encodeSigned" and logs the process.

- Response: Finally, it constructs a JSON response with the new token and sends it back. The response is of type "TokenResponse", containing fields "status" and "result".

This code provides an implementation for secure token management in a web service, ensuring that requests come from valid sessions or users by checking and issuing JWTs. It can serve as a starting point for you to implement both checking for tokens and issuing them. A simple middleware function that checks for token validity is the last in the file above:

```
jwtAuthMiddleware :: Middleware
jwtAuthMiddleware app req sendResponse = do
    let maybeToken = extractBearerToken req
    case maybeToken of
        Nothing -> sendResponse $ responseLBS status401 []
        "Unauthorized"
        Just token -> do
            let valid = decodeAndVerifySignature (JWT.toVerify
            . JWT.hmacSecret $ mainSecretKey) token
            case valid of
                Nothing -> sendResponse $ responseLBS status401
                [] "Invalid Token"
                Just _  -> app req sendResponse  -- Continue to
                the next middleware or main app
```

In order to use it, add it to the list of middleware already used in our src/WebAPI/WebApp.hs file in the main function that runs the Scotty server. However, you need to make sure you decide which routes should be open and which should be secured and adjust functionality accordingly, so that only secure routes are protected.

CHAPTER 13   AI MULTI-AGENTS, ARROWS, AND THE FUTURE

> **EXERCISE**
>
> Create a nice web API with protected and open routes. Open routes can, for instance, list available agents, while protected routes would require a JWT to access specific agents.

Error Handling: Embracing the Try–Catch Approach

Error handling is a topic that often turns the most seasoned programmers into keyboard warriors in defense of their favorite methods. In Haskell's land of lazy evaluation and type safety, dealing with errors may sometimes feel like traversing a labyrinth. But worry not! With the power of the *try-catch* approach, you too can elegantly manage those runtime surprises that life, or rather your code, throws at you.

Haskell primarily advocates for handling errors using types, leveraging constructs like `Maybe, Either`, and other monadic patterns – and we have seen lots of examples of their usage. However, there are times when we must descend from the lofty heights of type safety to the gritty world of exceptions – typically when interfacing with the real world, such as IO operations or dealing with foreign libraries.

Catching Exceptions: The Basics

At the heart of Haskell's exception handling lie functions from the **Control. Exception** module, particularly `try`, `catch`, and `handle`. These functions provide a way to deal with exceptions similar to the ubiquitous try–catch blocks found in imperative languages.

Let's dive into a simple example before we explore the deeper, uncharted waters. Look at the src/Errors/Sample.hs:

CHAPTER 13   AI MULTI-AGENTS, ARROWS, AND THE FUTURE

```haskell
import Control.Exception

testE1 :: IO ()
testE1 = do
    result <- try readNumbersFromFile :: IO (Either
    IOException [Int])
    case result of
        Left ex -> putStrLn $ "An error occurred: " ++ show ex
        Right numbers -> putStrLn $ "Read numbers: " ++
        show numbers

readNumbersFromFile :: IO [Int]
readNumbersFromFile = do
    content <- readFile "numbers"
    return (map read (lines content))
```

In this snippet, `try` is a heroic function that shields our `readNumbersFromFile` function from the ominous clutches of `IOException`. By wrapping the risky IO operation with `try`, we transform its type from `IO [Int]` to `IO (Either IOException [Int])`. The brilliance of this transformation is the return value now elegantly whispers into our control flow through pattern matching.

However, always remember: with great power comes great responsibility! The `try` function only catches exceptions that occur during the IO operation itself, not those that are hidden within lazy computations waiting to unleash their fury later.

Going the Extra Mile with Catch

For scenarios where you simply want to catch specific exceptions and handle them with grace, the `catch` function is your trusty sidekick:

```haskell
testE2 :: IO ()
testE2 = do
    catch (printDivisor 10 0) handler
```

```
printDivisor :: Integer -> Integer -> IO ()
printDivisor x y = do
    putStrLn $ "The divisor of " ++ show x ++ " by " ++ show y
    ++ " is " ++ show (x `div` y)

handler :: ArithException -> IO ()
handler ex = putStrLn $ "Caught an arithmetic exception: "
++ show ex
```

As demonstrated, `catch` takes two arguments: an IO action that might throw an exception and a function that knows exactly how to handle it – like a barista who already knows how you take your coffee when the coffee machine breaks down!

In this case, by catching `ArithException`, the program won't crash when our cheeky division by zero occurs. Instead, it calmly informs us of the wrongdoing with a comforting message, like a benevolent teacher guiding a math prodigy to avoid rookie mistakes.

Exception Handling with Style: The Handle Function

For those instances when you want a streamlined error-handling process without tangling your beautiful code with catch blocks, the `handle` function steps into the limelight. It lets you specify a handler just like `catch`, but it smuggles in the IO action implicitly:

```
testE3 :: IO ()
testE3 = handle handler $ printDivisor 10 0
```

Here, `handle` simplifies the syntax by taking the handler function first, followed by the IO action – perfect for those who appreciate a touch of syntactic sugar in their code diet.

CHAPTER 13   AI MULTI-AGENTS, ARROWS, AND THE FUTURE

While the try-catch methodology is not the default maneuver in Haskell's type-driven ecosystem, it harnesses immense power when dealing with the chaos of runtime exceptions. Whether you're catching specific exceptions or handling unexpected IO shocks, functions like try, catch, and handle ensure that you never lose your composure.

By embracing these tools, you can confidently weave robust and expressive error handling within the vibrant Haskell tapestry. Go forth, armed with this newfound knowledge, and build resilient applications that stay upright in the face of error winds and exception storms!

And of course it always makes sense to start with what we've invested so much into already – our AI framework.

EXERCISE

Think through where it makes sense to catch the exceptions in our AI framework code. MongoDB and API calls seem like good targets to check. Add the handlers to handle possible errors gracefully. Don't forget to log them properly through using our logger.

Multiple LLMs and Agents Support

Finally, it's worth making our AI framework as versatile as possible. For that, you can review the code in src/LLM/OpenAI.hs and create similar connections to other LLM vendors, such as Anthropic, Google, and others. Then you can create external integrations with other useful APIs such as web search or email access or CRM systems access – the sky is the limit. However, we will take a shortcut and integrate integrail.ai API. As I mentioned above, it not only gives you the ability to quickly create advanced AI multi-agents, supporting over 30 different LLMs as well as other GenAI models, but it gives you production Vector RAG out of the box, the ability to generate custom nodes to integrate with any custom

CHAPTER 13 AI MULTI-AGENTS, ARROWS, AND THE FUTURE

API, and much more. It is actually much easier to create the agents you need using Integrail and then use their power from your Haskell program – for instance, this way you can create a locally running agent that takes contextual information from your OS (which windows are open, what you are reading, etc.), and then you would pass it to the Integrail agent for processing and then perform certain actions on your system to assist you based on smart decisions made by AI. You are free to experiment as much as you like, and with the "magicalhaskell" code during registration, you receive $20 worth of credits for all range of external LLMs and other GenAI models.

Using integrail.ai agents locally from our framework is several easy steps:

- Register at integrail.ai.

- Copy one of the featured agents or build one from scratch by following documentation or tutorials.

- Upgrade your account to a FREE "Team" trial to get the $20 credits and be able to deploy your agents to the cloud.

- Once you pick an agent to deploy, simply click the "deploy" button (a little rocket) in the Agent Designer.

- Navigate to Settings of your account, go to the App Token tab, and get yourself a token – put it into a corresponding entry in the .env file of your framework code in this chapter.

- Finally, in the "Agents" tab, you should see something like Figure 13-13.

CHAPTER 13   AI MULTI-AGENTS, ARROWS, AND THE FUTURE

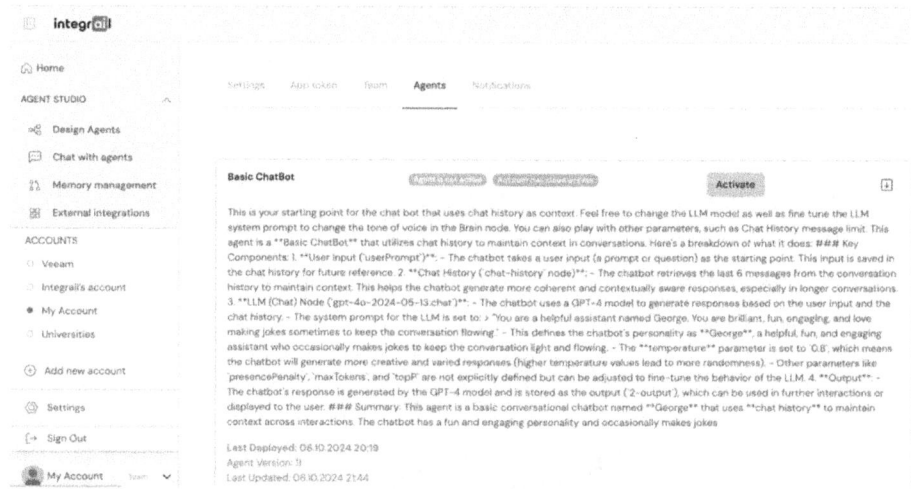

Figure 13-13. *Deployed agents at the Integrail cloud*

In the bottom of every agent, there's helpful information about how to call them using API, something along the lines of

```
curl -X POST https://cloud.integrail.ai/api/<your client id>
/agent/<agentId>/execute
-H "Authorization: Bearer <your token>"
-H "Content-Type: application/json"
-d '{ {"inputs":{"userPrompt":"what is (2+2)*2"}}}'
```

Note your client ID and the agent ID, along with the tokens, copy the new settings from the .env.template file to the .env, and fill in the necessary variables:

```
INTEGRAIL_TOKEN=your token
INTEGRAIL_CLIENT_ID=your id
INTEGRAIL_AGENTS=agentId1, agentId2
```

CHAPTER 13 AI MULTI-AGENTS, ARROWS, AND THE FUTURE

The INTEGRAIL_AGENTS setting lists the agent IDs that you want to work with from Jarvis. Integrail may expand the API to be able to list them automatically, but for now this is a small workaround.

Once you have done all this, you can check the src/Integrail/API.hs for the low-level API and additional commands we added to Jarvis in Middleware.hs and Commands.hs – but this code simply repeats all the patterns we have used many times throughout this book, so we are not going to review them here.

To test the agents, as always: stack build, stack exec jarvis, and – Figure 13-14.

```
Welcome to the Jarvis AI System!
Version 1.0.0

For help, type :help
In the multiline mode, type freely and then on the new line type :send to send the message
In the single-line mode, every time you hit 'Enter' the message will be sent
Enjoy!

[User]
:help
:send -- send a multiline message currently typed
:multi -- toggle multiline mode on or off
:embed -- create embeddings for the buffer text and save them to in-memory storage and mongo
:search n -- search in the Vector RAG and show top n items
:agent -- call integrail agent
[User]
hi
:agent
Ah, we're back to *hi*! 😊
At this rate, we're going start a world tour of greetings! Next stop: *Hello*, *Hey*, *Hola*, *Bonjour*... 🌍
```

Figure 13-14. *Jarvis is annoyed at me testing him with too many "hi's"*

Remember that even if code updates, you can always run :help in Jarvis, and it will show you all the commands available. :agent talks to the first available Integrail agent, but I'm pretty sure by the time the book is in print there will be a nicer interface to switch between them. Also don't forget to run tail -f .jarvis_logs in another terminal to monitor what's happening!

CHAPTER 13   AI MULTI-AGENTS, ARROWS, AND THE FUTURE

In any case, remember that the final, constantly updating, code for the whole Haskell AI framework that is loosely based on the concepts and code we created together in this book will always be available here: https://github.com/Integrail/magicalhaskell.

FINAL JARVIS EXERCISES

Here are some additional pointers for you in terms of how you can improve Jarvis and apply everything we've learned:

1. Graceful error handling and proper logging everywhere where it's needed

2. Additional UI commands to handle state, not just for the terminal interface, but for the web server, e.g., switching between agents or tracking the number of web connections to the server

3. Expanded web API with JWT-based authentication as suggested and described in the section above

4. "OS helper" agent that understands the current context in which you are working, sends it to an external AI agent, and automates certain tasks for you

5. Haskell programming copilot: agent that takes a user's input, generates code, receives compiler feedback, fixes code, rinse, and repeat until the code compiles

CHAPTER 13 AI MULTI-AGENTS, ARROWS, AND THE FUTURE

Conclusion

Haskell is an amazing, beautiful, and vast language. I hope that I succeeded to show you at least some sides of this beauty while making seemingly complex and confusing concepts easier to grasp. Please continue practicing and applying the functional typed way to solve programming problems:

- Understand the problem domain in terms of types.

- Decompose the problem into increasingly smaller pieces and come up with types and "arrows" between them, designing a small category of our problem domain.

- Look for abstractions along the way. Can this be made into a Functor? Is this a State monad? Maybe I should use Arrows?

- Visual representation always helps – I am sure I'd master Haskell much faster if somebody drew monad transformer stacks for me at least once.

- Apply all the spells and tools in our Haskell's magician toolbox to become fluent – type families? The ST monad for performance?

- Design the final application architecture in terms of a monad transformer stack or via Arrows. Design and start filling different "floors" – pure types and functions, IO low-level API, monads, lifts between them …

- Most of all, don't be afraid to experiment and learn new things!

CHAPTER 13   AI MULTI-AGENTS, ARROWS, AND THE FUTURE

Even though we covered lots of material, our main approach was to always discuss Haskell through the lens of types and relations between them, kind of a categorical approach. There are so many more things we didn't touch – existential types, proper use for rank-N polymorphism, dependent types (very actively researched in Haskell), etc. I encourage you to explore on your own.

As for AI, this is undoubtedly a significant revolution that's coming to all sides of human lives. Code generation among other things is one of the primary tasks for many AI startups and researchers – and even with Jarvis we have created, you can get quite a bit of Haskell help by providing enough context of your code and imported libraries. AI can help us a lot in mundane, day-to-day tasks – e.g., it generates ReactJS UI code extremely well even today. However, what AI can't do and probably won't be able to at least in the medium-term perspective is invent truly new things. And if anything, Haskell is one of the best languages to continue inventing new things and in turn enrich AI agents so that we outsource the mundane and focus on creative.

Not just that, but also as we have tried to show in this chapter, Haskell is so well suited with the Arrows approach to model various distributed execution pipelines, which perfectly fit into any Agentic AI framework you can come up with. If you are passionate about Haskell, if I piqued your interest even a bit, let's work together to make this wonderful language even more popular and more used in practice.

If you are thinking "in which areas is Haskell currently lacking," I would name but one: good GUI. There are many different libraries that sort of work or don't work, but there is no standard, nice, easy way to create GUI applications in Haskell – be it "desktop" or on the Web, which is a pity, since again, the expressiveness and robustness of the language could improve the quality of software around us significantly. So if you would like to contribute to the community, this, along with AI, is a good place to start.

I hope you enjoyed this journey as much as I did, and if you have questions, comments, or ideas, you are always welcome to write to me and I will do my best to reply to everyone.

Index

A

Aeson library, 228
AI agents framework
    agent functionality, 303
    MRWST monad
        transformer, 314–320
AI chat agent
    chat history, 255–257, 259–262
    Jarvis, 237
    logging function, 248–251, 253
    terminal-based UI
        business logic, 239
        initialization, 241, 242
        Jarvis, 239
        main function, 245–247
        monad transformer stack,
            242, 244, 245
        response function, 239, 240
    writer monad, 253–255
AI multi-agents
    error handling
        catching exceptions, 396–398
        style, 398
    GUI, 405
    LLMs, 399–402
    web authentication, JWTs,
        390–392, 394, 395

Algebraic data types (ADTs), 25, 57
anotherWeirdFunction, 17
Applicative functors, 114
Arrows
    AI multi-agents
        ChatGPT, 374
        creation, 379–383, 385–389
        examples, 377–379
        LLM-based module, 375
        OpenAI LLM, 373
        URLs, 375
    definitions/examples, 364–367
    interface, 367, 369–373
    monad transformer stacks,
        361, 362
    typeclasses, 361
    typed functional
        programming, 363
Artificial intelligence (AI)
    Haskell/LLM
        framework, 211
        MongoDB, 210
        terminal chatbot, 212–214,
            217, 218
    LLM
        ANNs, 204
        fine tuning, 209

INDEX

Artificial intelligence (AI) (*cont.*)
    limitations, 207
    text generation, 207
    tokens, 205
    vector embeddings, 206
    vector RAG, 208
Artificial neural networks (ANN), 203
Associated type and data families, 328

B

Bifunctor, 114
    applicative typeclass, 99–101, 103, 105–107, 109–111
    bimap, 96
    generic function, 94
    generic type, 97
    Maybe, 93, 102
    PlayerData type, 103, 104
    regular functions, 98
    String Double, 93
Blackjack program
    domain modeling, 201
    final computation, game cycle, 197–200
    GameFloor, create file, 188–195, 197
    implementation, 171
    IO, 183–186
    iterative development and testing, 202
    layered architecture, 201
    "main" function, 171
    preparation, 174–176
    pure design, 177–183
    solve problems, 172
Boxed vectors, 331

C

cabal and stack for package management, 174
"Call by need" lazy evaluation approach, 278
Cards.hs, 177
Cartesian product, 61
Category laws, 364
Chatbots, 204, 209, 263, 360
chatCompletion function, 228, 245
ChatGPT, 204
"C--" language, 279
Comprehension, 46
Constructor function, 55
Coproduct types, 61, 116
Currying, 17

D

DeckM, 122, 126, 132, 151
Deep Reinforcement Learning, 204
Dependent function type, 59
Dependent pair/sigma types, 62
Dependent types, 51, 58
Do notation, 138, 153, 186

E

Endomorphism, 68
execStateT function, 193, 198
extractBearerToken
    function, 394

F

findAll method, 348
Full-featured terminal-based
    application, 237
Functional programming, 15
Functional dependencies, 132
Functional reactive programming
    (FRP), 360
Functor, 36, 82, 83, 113, 359, 404
    three-dimensional vector
        example, 89–91
    tracking players, game, 91, 92
    typeclass, 114
    typeclass definition, 86–88

G

GameFloor, 188
GenAI models, 400
Generalized algebraic data type
    (GADT), 31

H

Hask, 8
Haskell's aeson library, 235

I

Identity, 160
id function, 383
incrementVector
    function, 333
initAll function, 242
Input output (IO) monad
    Blackjack game, 159, 162
    computations, 148, 153
    DeckM, 150, 151
    designers, 147
    do notation, 153, 155, 156
    effects, 164, 166, 168
    Haskell code, 152
    Haskell libraries, 153
    Haskell programs, 169
    identity, 160
    imperative language, 149
    lift function, 161, 163
    operations, 167
    single monad, 158
    state, 160
    typeclass, 150
    typical real-world
        program, 163
    "where" block, 152
InputT monad transformer, 238
IOFloor.hs, 184

J

Jarvis, 265
JSON Web Tokens (JWTs), 390

INDEX

K
Kind system, 54
Kleisli constructor, 372

L
Lambda function, 240
Large language models (LLMs), 204, 234
List, 40
ListInt, 16, 19

M
Magma, 62, 74, 83
Map, 16
mapM_C function, 259
Maybe type function, 83
Monads, 88
    applicative, 116, 117
    card playing
        competition, 117–128
    countable, 117
    reader, 135–139
    state, 132–134, 145
    transformer stacks, 146
    typeclass, 114
    typeclass/basic
        monads, 128–131
    writer, 140–145
Monad transformer stacks, 114
MongoDB, 211, 303–305, 360
    initialization, 308, 309
    no-SQL database, 305–307
    system prompts collection, 310, 312, 313
    terminal and web server modes, 308
    web API functionality, 304
Monoids, 75, 113
Morphisms/maps, 63
Mutable variables
    relationship, IORef/MVar/TVar, 299
    functional programming, 295
    IORef, 295, 296
    MVar, 296, 297, 300
    performant algorithms, 294
    TVar, 298–300

N
Nat, 52, 74
Normal functions, 367

O
OpenAI, 204, 360
OpenAI's API, 238
    chatCompletion function, 228, 230
    chatoptions, 220, 222
    handling optional fields, 226, 227
    Haskell, 219
    JSON, 222–225
    message type, 220
    program, 231–233

streaming fashion, 230
types/JSON, 219

P

Parametric types, 72
Partial function, 10
Perceptron, 203
Polymorphic function, 43
Prelude, 18
Primitive types, 281
Primops/primitive operations, 17
Product types, 24

Q

Query, 378

R

Rank-N polymorphism, 405
Rank-N types, 360

S

Schema data type, 312
Sigma types, 52, 72, 325
Software Transactional Memory (STM), 298, 334
Spineless Tagless G-machine (STG), 278
State monad, 121, 126, 151, 359
Structure-preserving maps, 87
System FC, 51

T

Tokens, 205
Transformer stacks, *see* Input output (IO) monad
Traversable typeclass, 116
Type/category theory
    advanced generalized functions, 54–58
    characteristics, 63
    dependent function type, 59, 60
    functional programming, 63
    Haskell, 51
    morphisms, 64, 65
    sum/product/dependent pair types, 61, 62
    typeclasses, 66–69
    types and functions, 52–54
Typechecker, 3, 49
Typeclasses, 49, 50, 69, 85
    algebra, 74, 75
    Char or Int, 71
    example, 73
    functor, 82
    Magma, 83
    Maybe/list, 79–81
    monoid, 76–78
    multiparameter, 72
    show, 73
Type construction
    ADTs, 25, 26
    data constructors/types, 37, 38
    deck of cards, create, 21, 22
    functions, 30–36

Type construction (*cont.*)
    house of cards, algebra, 22–24
    list, 40–43
    map function, list, 44, 46–49
    records, 27–30
Typed functional programming, 3
Type families
    associated types/data, 325–330
    data, 323, 324
    functions, 321, 322
    synonym, 322
Type functions, 32, 54, 72, 364
Type-level functions, 14
Types
    functions, 8
        char/int, 8–12
        curry/recurse, 15–18
    string, 12–14
    synonym, 48

U
Unboxed vectors, 331

V
Values, 19
Vector embeddings, 205
Vector library
    boxed and unboxed vectors, 331
    implementation/generic
        interface, 331, 332
    mutable vectors/ST
        monad, 332–337
    practice, 339, 340
    rank-N types, 337–339
Vector RAG, 304, 341, 360
    implementation, 342, 356
    in-memory vector
        search engine
        implement
            functions, 352–355
        Mongo storage, 344–348
        store, 348–350
    terminal and web UI, 340
    testing Jarvis, 357
    text-to-vector
        embeddings, 342
verySafeDivide, 94

W, X, Y, Z
Web-enabled AI
    GHC
        adding web API,
            283–289, 291–294
        Jarvis program, 265
        newtype keyword,
            280–282
        STG, 278, 279
        terminal UI, 266–275
Wizards
    problem, 1, 2
    solving problems, 3–7

GPSR Compliance

The European Union's (EU) General Product Safety Regulation (GPSR) is a set of rules that requires consumer products to be safe and our obligations to ensure this.

If you have any concerns about our products, you can contact us on

ProductSafety@springernature.com

In case Publisher is established outside the EU, the EU authorized representative is:

Springer Nature Customer Service Center GmbH
Europaplatz 3
69115 Heidelberg, Germany

www.ingramcontent.com/pod-product-compliance
Lightning Source LLC
LaVergne TN
LVHW010333260326
834688LV00036B/691